Generations of Data Warehousing and Business Intelligence Technologies and Solutions

I0464538

FIRST EDITION

Author
Venkata Surya Brahma Linga Sarma Somina
Senior Principal Architect Data Warehousing and Business Intelligence

Copyright © 2012

Hard Book:
ISBN:13-9781514603581(USA)
ISBN:10-1514603586(Rest of the World)
Amazon Kindle Book: ISBN: 9781973550426

Publisher
Amazon Create Space Print

Preface from first edition

PREFACE

This is amazing and very rare attempt to have content of the book like this. I have been in the industry for quite a long time, I have seen many people have different kinds of opinions the way organizations function in a day to day life. Evolving around more maturity and very intelligent way organizing information one of the KPI for the data warehouse implementations and features. Many Data warehouse are functional and own the business transformation and knowledge. But, many clients have challenges with respect to maintaining their data warehouses Growing volumes and variety of data has given more operational insights for the data warehouses. Data cleansing and other areas like handling various formats of data is another challenge. I my-self noticed, after spending considerable period of time and had hands-on on many of these data warehouses, there could be real way of looking into various generations, how they have been transformed. It took more time than required for the industry to evaluate the need for the new data warehouse systems having more disciplined approach. Cloud acts as one of the latest generations which build data warehouses with intuition level data. Bigdata has more to do with social media and web traffic analytics with analytical services and appliance engines. I have attended many seminars, conferences organized by TDWI, Ralph Kimball, Bill Inman and Global STF, everyone has different opinion and approach on all the data warehouses and business intelligence solutions are built. The databases and database theory have been there for a long time. Earlier database solutions were centered on a single database serving all the needs of information processing from OLTP to OLAP solutions. The primary focus of the early database systems were operational with transactional systems like OLTP. The initial stages of DWH and BI solutions are built just to get the data from these transactional systems for reporting purpose. It has emerged to a level where the information needs to be processed is of historical and should be maintained to have a data of years together. The generations of technologies and solutions illustrated in the book are to support better MIS and analytical platforms.

Over a period of time, data warehousing has surpassed the database theoreticians who wanted to put all the data in a single database. Data warehousing survived the dot.com disaster brought-on for a quick money making solutions. Today the data warehousing is not theory it is fact of life. The cost of information has dropped dramatically because of the data warehouses. Through the evolution of various data warehousing solutions, the integration of data across the corporation has become a real possibility, in most cases for the first time. Believe that data warehousing is a technology not an information alone that confined to a specific solution. There is a relationship between data warehousing and database technology. It differs in information processing shows that such as its tendency to dwell on detail based on the data warehousing and database technologies. There is a notion that if we get results right the end result will somehow take care of itself and we will achieve the success. There are four levels of data and processing in the architecture – the operational level, the data warehouse level, the departmental/data mart level and the individual level. Most of the conventional applications have been developed by SDLC life cycle, but the data warehouses were developed under an approach called the spiral development methodology and have interfaces to object oriented programming languages and Agile BI. If you're a data warehouse and Business Intelligence professional you would have probably noticed, many new options for data warehouse platforms that have appeared for the last few decades. We've seen the emergence of new categories of data warehouse (DW) platforms, such as data warehouse appliances and software appliances. A real time data warehousing is there in the market years together. A new interest in columnar databases has led to several new vendor products and renewed interest from the older ones. Open source Linux is now common in data warehousing and open source databases, data integration tools, and reporting platforms have given an opportunity to look into more investments. In the hardware realm, 64-bit computing has enabled larger in-memory data caches. Leading database vendors have added more features and products conducive to data warehousing. The evolution of the new data warehouse appliances and advanced analytics facilitates faster and ease to implement in-built platform and user defined functions and integrated an in-database analytics framework. Furthermore, a number of data warehouse platforms and other business intelligence platforms are now available through software-as-a-service (SaaS) and cloud computing. Mobile BI is another adoption and evolution of the capability that enables to gain business insights through information analysis using applications optimized for mobile devices like i-phones and tablets with reduced development life cycle. Data Science and Scientists have challenges to invent and integrate them with the existing data warehouses to make them operational. The focus is more on technology, but this book also explains how technology's adoption in next generation data warehouse platforms is driven by real-world business and organizational needs, requirements, technology challenges, data formats and cost.

This book is about the evolution of data warehousing and business intelligence technologies and solutions. The discussions in this book are geared to all the Architects and Consultants working in Data Warehousing and Business Intelligence space. The industry may get assisted from the frame works that were developed part of this book are comparable to understand the benefits of each generation of Technology and Solution. This will help the industry to focus up-on building new information together for the best adoption and to understand the major areas of concern. This book is also to help the computer professionals of the new era, how the systems are developed and transformed to meet the latest requirements of data processing. It serves the purpose, as the book for reference and guide for data warehousing and business intelligence solutions.

I am dedicating this book to clients, colleague and expats are in the industry, particularly in data warehousing and business intelligence space and to my boy who has given time, has sustained me through all the ups and downs in life.

Table of Contents

CHAPTER 1
INTRODUCTION

What is new generation of data warehousing?

Data Warehousing and Business Intelligence literature provides certain relatively new technologies, techniques, and business driving the practices majority of data warehouses and their platforms towards a redesign, major retrofit, or even replacement that we can recognize as a generation. The generations change is an evolutionary process that adopts the resulting data warehouse to changing business and technology requirements to facilitate the research and to growing demand for cost reductions and opportunities and to accommodate growing volumes of data.

We need to adopt the systems of new era to understand the information technology to a possible solution to resolve the known database problems. Many of the problems are related to storing of information in a right format and present it in right way where ever necessary. The data storage has become a challenge, where any particular format is not able to support, thus become a harm string approach for the evolution of new data warehouse/BI generations and provide support for any specific notation. We have any model to adapt to need for answering the queries of business. The professionals have roots can be traced back to antiquity. The more fundamental way we evolve is to adapt to a specific modeling technic and to follow the basis of information management. The research found that the historical data of all systems in the organization may not be very relevant to the business. In some cases it has to follow a pattern based approach to resolve more issues related to information management and to support data issues.

Historically, the first generation of information era not focused much on these techniques of information storage either due to lack of knowledge or not to have servers/hardware to support these. The evolution information technology in various technology spaces including web technologies and also tremendous advancements in the hardware side, increased consumer based sales to have more transaction based capability has given an insight for the database systems experts to look into areas of development. It has also become a challenge for the transaction based systems to handle on- going concurrent transactions of business to make it optimum level of through-put. There was a tremendous demand for the campaign management, vendor management solutions, which in fact gained more momentum in early 1990s.

The first generation of data warehousing and business intelligence was only focused on simple data storage, thus facilitate only representation of data. Defining the granularity of data has become biggest challenge for the designers. Growing demand based business and to reduce the cost on has helped the organizations to have multiple storage systems across the globe. Interestingly, there is more focus on sales and pattern based sales, customer based sales, develop value added data sets, geographic based sales all these have come into picture to improve overall revenues for the organizations and to facilitate the fast information access. Thus information era was more automated to understand increased revenues and cut-down the costs to a level of obtaining maximized profits for the organizations. Any particular domain specific business has to evolve around well-developed solutions, thus make it easy for the business to run their day-to-day operations. Generations of various technologies have evolved to produce the competent models, to increase and to accommodate the growing demand and solve many issues of operational.

Federated way doing business is thus evolved in the next generation to integrate the desperate database systems late 1990s. This has facilitated accessibility to data in a deferent region to understand their sales patterns. More or less, most of business is same, to an extent of consumer based business. The federated systems have issues with data synchronization and other issues of connectivity, provide better accessibility of date in different region. In some areas, there are data latency issues, not able to gain the clear picture of what business is. Accordingly, there is more demand for the consumer based models, to understand growth patterns of revenues. In-tandem research has looked into these issues in early 21st century, to address the problems of the data accessibility issues.

Real time data warehousing and actionable data warehouse systems have thus evolved in early 2003. This has reduced the demand for model based approach of storing data. Many new applications are developed to promote and protect the business and the customers to follow the regulations and fraud detection. Also to enable transaction based systems to understand these new developments in multiple application areas. There is huge demand for data on-time techniques. Founded there are information storage is required to support these applications and to have more automation of information. Moreover, all this information should be accurate enough to produce the right results, not to have any discrepancy on the information storage and activity monitoring has to keep the business growing. This has become a more tedious process for the new systems to connect together.

This did not help the business much it has added huge cost in terms developing new applications, new generation of data warehousing looked at in and about 2005 to develop systems to have more data management practices. Master Data Management has thus evolved to resolve the issues of data cleansing related to de-duplication and redundant information storage etc., to growing demand to have single customer view. This has facilitated to have more disciplined approach to have single customer-view of information conformed way, to obtain more value added services to the customers. This improved the business analytics models to present the right information, to support loyalty based business to have more centralization of information to reduce the costs of information storage and to promote security issues.

Research had looked into other areas to promote the 64-bit computing, instead 32-bit computing to store more information on the same node to promote Service Oriented Architecture (SOA) for faster processing. Web based technologies simultaneously developed to promote and use the 64-bit computing in multi-tier platforms. The growing demand and also to facilitate more and more technology based solutions 64-bit computing has found more value on all most all the database vendor technology to have processing power of mainframes and legacy systems. The value addition here is to have faster information access, storage, performance and reliability and provide support for CPU based processing of data. There are still data storage issues.

Many vendors have looked into open source systems and freeware to reduce the cost on the database management and to reduce the multiple vendor dependencies and demand.

New database vendors have come into market to reduce and cut-down the cost for enterprises. New vendor management and open sources have challenged the old era of databases systems. Many of the solutions have developed to have domain specific models and applications, in-order to reduce the cost of multiplied applications development. Over a period of time, the cost on the systems should reduce dramatically. This has led to more competition on keeping the customers in tact with business and also to have object-oriented UML based solutions. It is there-fore obtained to have consistent approach in all these solutions to understand the process of knowledge based solutions not to have discussions and hence forth to have technology based approach to all these solutions across the world. Many database vendors have come into the market providing free-ware solutions and thus formed a network of database management solutions to have a congestive approach.

Data warehouse appliances are ready to use solutions. Latest trends like cloud computing, 64-bit computing, columnar databases and data warehouse appliances and software appliances have provided new insight and demand for the latest technologies all together. Apart from the technology revolution, these latest advanced platforms also provide support for the on-going demand and scalability. There is more demand for less cost implementation and product based technologies, where adapting to the new platform is easy, quickly understandable and user-friendly. All these up-course provides lot of encouragement for venturing into new technologies and platforms. There are other challenges like problems of sustainability, there could be limitations, with respect to data storage, performance, like you see, many vendors still running systems with legacy and mainframe platforms of old era.

TDWI periodically organize for many surveys interns of introducing new vendors, business user adaptation to technology change and also the open source platform management. Keeping these in view TDWI understands that these requirements and priorities will assist users as they plan their next generation data warehouse platforms.

1.1 Technology Drivers for the evolution of New Generations of Data Warehouses

There's more than one path to the next generation of your organization's data warehouse platform:

Retain the current platform, but do more with it – It is a kind of enhancement approach evolve around developing of the systems on existing ones, thus retain the current platform. There is a difference between a data warehouse and the platform that manages it. The re-modeling of the DW is significantly to add value without replacing the platform that manages it. Incremental additions to hardware are common (to add more CPUs, memory, or storage), and these satisfy next generation requirements for fast queries, in –memory databases and scalability, by doing more with the current platform.

Replace the current system; build the new one and encapsulate -This is more advanced approach in maintaining the data warehouse systems update-to-date with the technology and to have the access to latest technologies and trends, depending on the cost and ease of implementation and risk. Building completely new platform is time spending, but there is always need for the new IT experts and research to do this for the said benefits

Generational DW Features	Using now(%)	Use 3 years(%)	Plan to Use(%)
Real Time Data Warehousing	60	80	60
Master Data Management(MDM)	30	40	50
On-premise cloud	20	50	50
Hybrid cloud	40	60	60
public cloud	40	20	20
Private cloud	40	20	20
Software-as-service Analytical	30	60	80
In memory databases	25	25	25
Open Source DBMS	32	60	60
Service Oriented Architecture(SOA)	35	50	50
Advanced Analytics (mining, predictive)	35	50	50
Data Warehouse Appliances and Power Efficient Hardware	30	70	70
MPP	40	40	50
Web Services	40	70	70
Software Appliances	40	20	20
High Availability systems	40	70	70
Column Oriented Storage Engines	50	30	30
Advanced Analytics (Presciptive)	40	70	70
CDW and EDW	60	50	50
Atomic Data warehouses	40	50	50
Data Lakes	30	70	70
Data encryption and security features	50	70	70
Cyber Security and Identity management	30	50	60
Data Visualization	50	40	30
Self Service BI and Agile BI	20	60	60

Figure 1.1 - Illustration of Generational DW Features or Technique

The above is the illustration of the business growth of various technologies and platforms as on today and going to be. As per Gartner Analysis, the traditional platforms would continue to survive, as support costs are going to play major role and also accommodate the new technology skill development and other areas.

CHAPTER 2

EVOLUTION OF DECISION SUPPORT SYSTEMS

A Decision Support System (DSS) is a computer-based information that supports business or organizational decision making activities. DSSs serve the management, operations, and planning levels of an organization (usually mid and higher management) and help people make decisions about problems that may be rapidly changing and not easily specified in advance - i.e. Unstructured and Semi-Structured decision problems. Decision support systems can be either fully computerized, human-powered or a combination of both. While academics have perceived DSS as a tool to support and decision making process, DSS users see DSS as a tool to facilitate organizational processes.

Information Systems researchers and technologists have built and investigated computerized Decision Support Systems (DSS) for approximately 40 years from now. The journey begins with building model-driven DSS in the late 1960s, theory developments in the 1970s, and implementation of financial planning systems, spreadsheet-based DSS and Group DSS in the early and mid-1980s. Data warehouses, Executive Information Systems, OLAP and Business Intelligence evolved in the late 1980s and early 1990s. Finally, the chronicle ends with knowledge-driven DSS and the implementation of Web-based DSS beginning in the mid-1990s. The field of computerized decision support is expanding to use new technologies and to create new applications.

- DSS tends to be aimed at the less well structured, underspecified problem that upper level managers typically face;
- DSS attempts to combine the use of models or analytic techniques with traditional data access and retrieval functions;
- DSS specifically focuses on features which make them easy to use by non-computer people in an interactive mode; and
- DSS emphasizes flexibility and adaptability to accommodate changes in the environment and the decision making approach of the user.

The evolution of DSS system is phenomenal and orchestrated with growth of data through what has happened to predicting what will happen. The need and demand building data warehouse systems surpassed the earlier version of just information storage and querying. The data Warehouse requires architecture that begins by looking at the whole and then works down to the particulars.

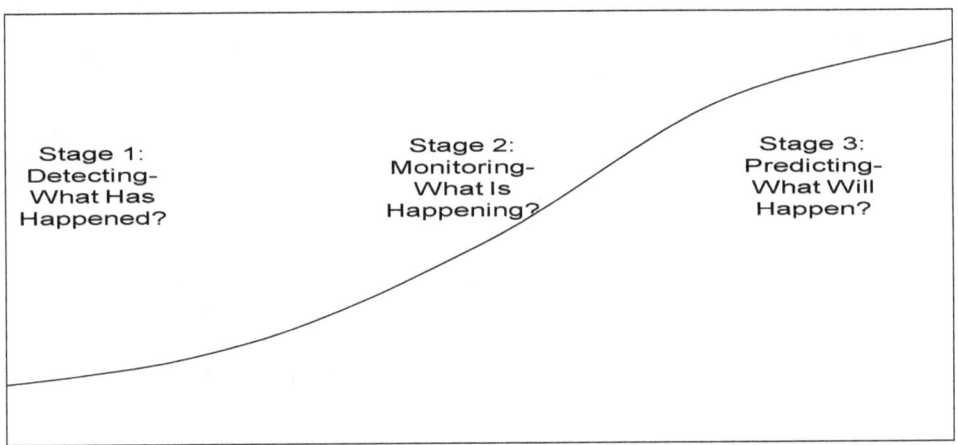

Stage 1:
Detecting–
What Has
Happened?

Stage 2:
Monitoring–
What Is
Happening?

Stage 3:
Predicting–
What Will
Happen?

Figure 2.1 – Decision Understanding Stages

As the emerging trends and technologies introduced the problems with productivity with many data warehouse systems come to an examination of review found that the data credibility is the only major problem with the naturally evolving architectures and frameworks. In order to locate the data many files and layouts of data must be analysed. In earlier versions some files used Virtual Storage Access Method (VSAM), and some use Information Management System (IMS) and some use integrated database management systems (IDMS).

The next task is to produce the report to compile the data once it is located. The transformation from data to information as the existing applications simply do not have the historical data required convert the data to information. A change approach thus evolved to build an architected data warehouse to store the primitive and derived data. Derived data is part of the DSS systems summarized or otherwise calculated to meet the business needs of the organization. The architected environment for building the data warehouse has natural extension of the split in data caused by the difference between primitive and derived data.

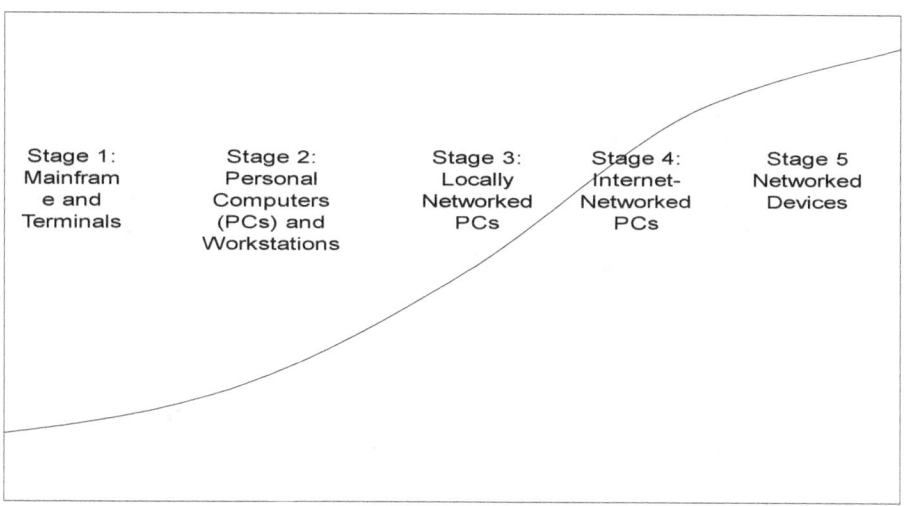

Figure 2.2 – DSS Hardware Evolution

Typical information that a decision support application gathers and present includes:

- Inventories of information assets (including legacy and relational data sources, cubes, data warehouses and data marts),
- Comparative sales figures between one period and the next, projected revenue figures based on product sales assumptions.

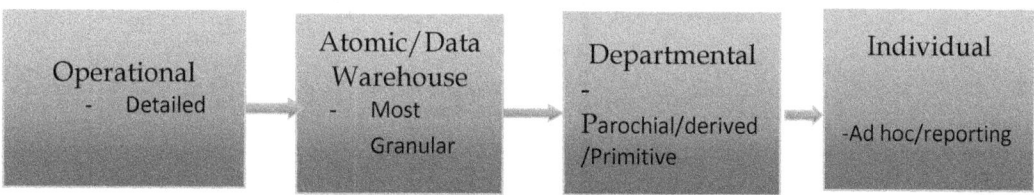

Figure 2.3 – Levels of the data Warehouse Architecture

2.1 Granularity of the data warehouse

The single most important design issue any data warehouse is the challenge to identify the granularity. While designing the DW/BI application it is essential we identify the granularity at the right level will facilitate the smooth implementation. The primary issue of granularity set to the level of defining drill-down and roll-up nature of the database on level of detail of information required. In some cases it is very detailed like customer information of CDRs, itemized calls and monthly bank statements, in some cases the detailed to a level of GDP and Union budgets, per capita income, etc. Also to define to a level one service provider or for that matter a defined rate plan revenues on cumulative bases. All this information needs to be designed and loaded to the database systems, where the reporting layers can access this information present the information in a required format.

2.2 Benefits of Data Warehousing Granularity

As said above, thus facilitates on demand for the presentation of the summary as well as detailed. A well designed approach of granularity will reduce data cleansing life cycle and also provide to memory efficient database storage. Apart from this there are also benefits hierarchy building, enhanced data provisions and any extrapolation required. Other meaning of the granularity is the flexibility the way it can adopt to multiple scenarios.

- Types of Granularity
- Role-up models
- Drill down Models
- Custom Hierarchy Building

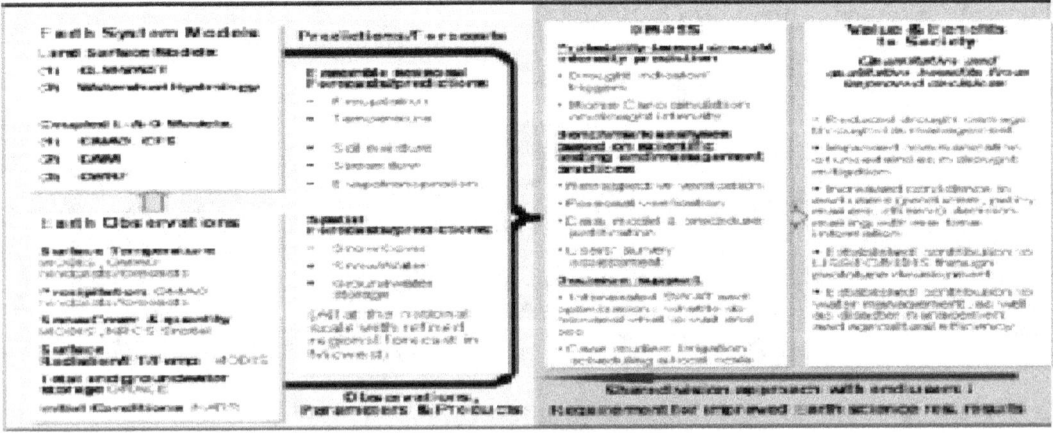

Figure 2.4 - Components of DSS system

Three fundamental components of a DSS architecture are:

- The database (or knowledge base),
- the model (i.e., the decision context and user criteria)
- the use interface

2.3 DSS Frameworks

DSS systems are not entirely different from other systems and require a structured approach. Such a framework includes people, technology, and the development approach.

- o Intelligence - Searching for conditions that call for decision.
- o Design - Developing and analyzing possible alternative actions of solution.
- o Choice - Selecting a course of action among those.
- o Implementation - Adopting the selected course of action in decision situation

DSS technology levels (of hardware and software) may include:

- The actual application that will be used by the user. This is the part of the application that allows the decision maker to make decisions in a particular problem area. The user can act upon that particular problem
- Generator contains Hardware/software environment that allows people to easily develop specific DSS applications
- Tools include lower level hardware/software. DSS generators including special languages, function libraries and linking modules

2.4 DSS Application Development

2.4.1 Model-driven DSS

A model-driven DSS emphasizes access to and manipulation of financial, optimization and/or simulation models. Simple quantitative models provide the most elementary level of functionality. Model-driven DSS use limited data and parameters provided by decision makers to aid decision makers in analyzing a situation, but in general large data bases are not needed for model-driven DSS follows more of a relational data base model.

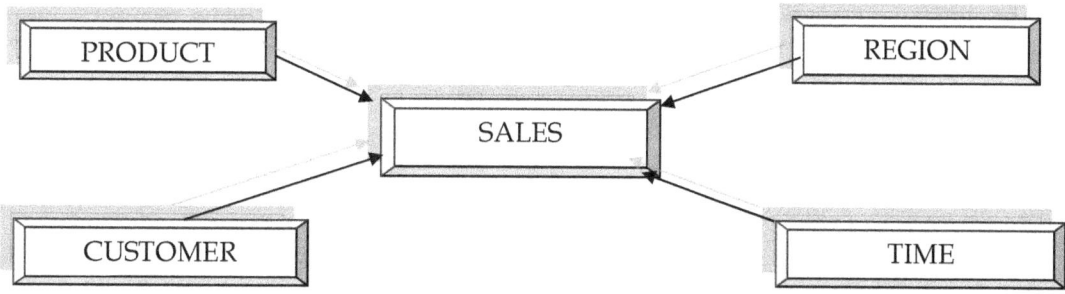

Figure 2.4.1 – Model Driven DSS

2.4.2 Data-driven DSS

A data-driven DSS emphasizes access to and manipulation of a time-series of internal company data and sometimes external and real-time data. Data-Driven DSS with On-line Analytical Processing provide the highest level of functionality and decision support that is linked to analysis of large collections of historical data. Most data-driven DSSs are targeted at managers, staff and also product/service suppliers. It is used to query a database or data warehouse to seek specific answers for specific purposes. Executive Information Systems (EIS) and Geographic Information Systems (GIS) are special purpose Data-Driven DSS. On-line Analytical Processing (OLAP) software is used for manipulating data from a variety of sources that has been stored in a static data warehouse. The software can create various views and representations of the data. Executive Information Systems (EIS) are computerized systems intended to provide current and appropriate information to support executive decision making for managers using a networked workstation. EIS offer strong reporting and drill-down capabilities. A Geographic Information System (GIS) or Spatial DSS is a support system that represents data using maps. It helps people access, display and analyze data that have geographic content and meaning.

Figure 2.4.2 – Data Driven DSS layers

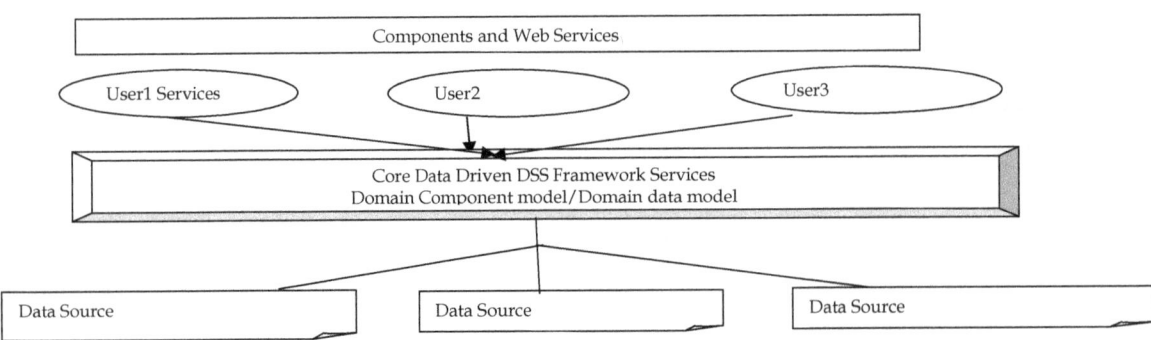

Figure 2.4.3 – Data Driven DSS Framework

2.4.3 Communications-driven DSS

Communications-driven DSS use network and communications technologies to facilitate decision-relevant collaboration and communication. In these systems, communication technologies are the dominant architectural component. Tools used include groupware, video conferencing and computer-based bulletin boards. In addition to the primary technologies for communications-driven decision support, voice and video delivered using the Internet protocol have greatly expanded the possibilities for synchronous communications-driven DSS. Communication driven DSS software have one of the following characteristics:

- Enables communication between groups of people
- Facilitates the sharing of information
- Supports collaboration and coordination between people
- Supports group decision tasks

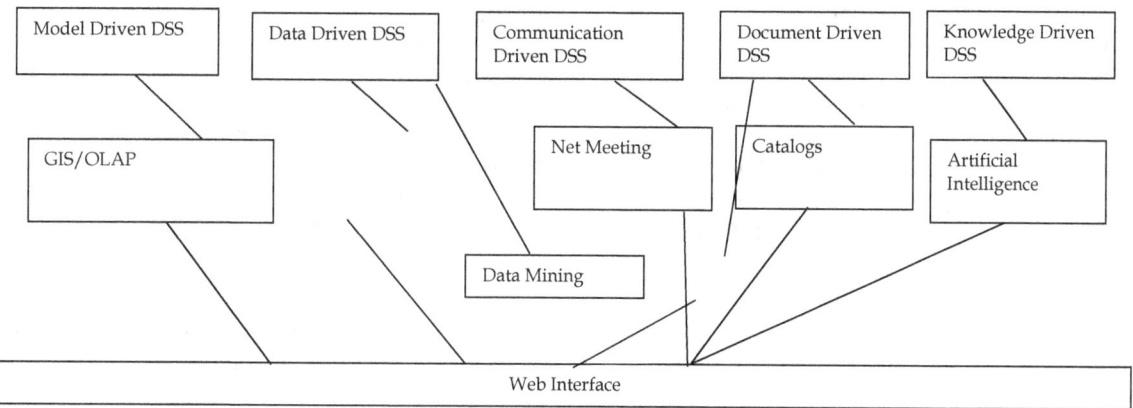

Figure 2.4.3 – Communication Driven DSS framework

2.4.4 Document-driven DSS

A document-driven DSS uses computer storage and processing technologies to provide document retrieval and analysis. Large document databases may include scanned documents, hypertext documents, images, sounds and video. Examples of documents that might be accessed by a document-driven DSS are policies and procedures, product specifications, catalogs, and corporate historical documents, including minutes of meetings and correspondence. A search engine is a primary decision-aiding tool associated with a document-driven DSS. These systems have also been called text-oriented DSS. The World-wide web technologies significantly increased the availability of documents and facilitated the development of document-driven DSS.

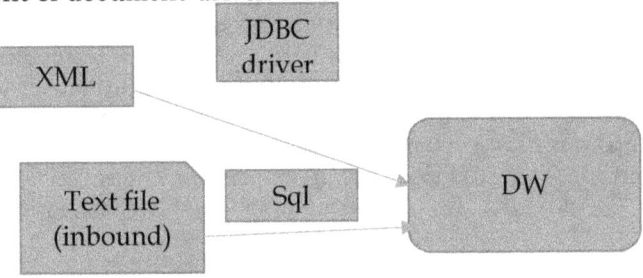

2.4.5 Knowledge-driven DSS

Knowledge-driven DSS can suggest or recommend actions to managers. These DSS are person-computer systems with specialized problem-solving expertise. The "expertise" consists of knowledge about a particular domain, understanding of problems within that domain, and "skill" at solving some of these problems. These systems have been called suggestion DSS and knowledge-based DSS examined Artificial Intelligence (AI) contributions to DSS. Artificial Intelligence systems have been developed to detect fraud and expedite financial transactions, many additional medical diagnostic systems have been based on AI, and expert systems have been used for scheduling in manufacturing operation and web-based advisory systems. In recent years, connecting expert systems technologies to relational databases with web-based front ends has broadened the deployment and use of knowledge-driven DSS.

2.4.6 Web-based DSS

The World-wide Web and global Internet provided a technology platform for further extending the capabilities and deployment of computerized decision support. The release of the HTML 2.0 specifications with form tags and tables was a turning point in the development of web-based DSS, In addition to Web-based, model-driven DSS, reporting Web access to data warehouses. More sophisticated "enterprise knowledge portals" are introduced by vendors that combined information portals, knowledge management, business intelligence, and communications-driven DSS in an integrated Web environment.

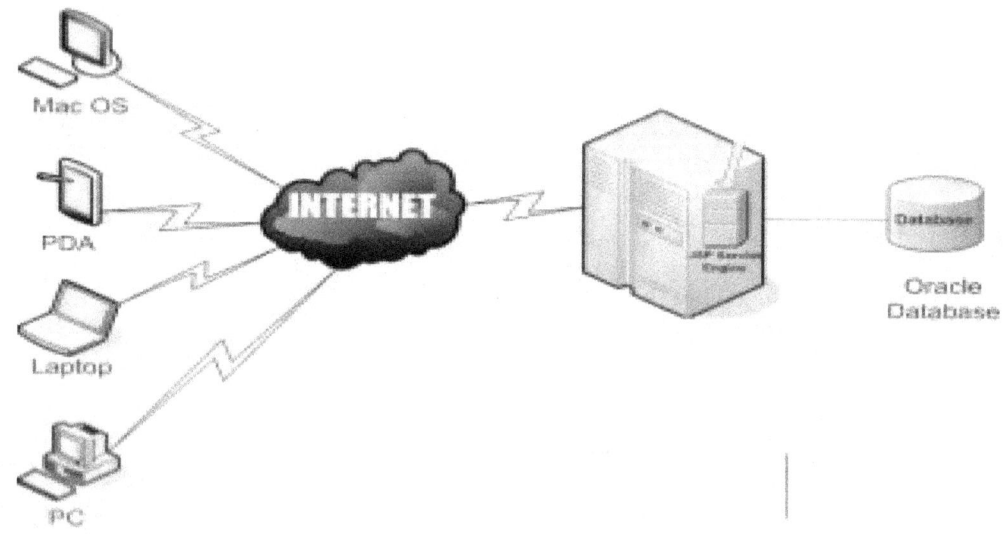

Figure 2.4.6 – Web-based DSS

CHAPTER 3
EVOLUTION OF ODS AND OLAP

In the early 2000s, the Gartner Group found the real-time enterprise (RTE). HP was a leader in architecture centered on an operational data store (ODS) that was similar to the data store used by an online data warehouse. Rather than periodically loading the ODS with massive amounts of data via an ETL facility, the ODS is trickle-fed transactions as they occurred so that it always contained the latest state of the enterprise as well as historical information. Using the ODS, classical data-mining engines, generate strategic information and knowledge, and real-time rules engines could make tactical decisions regarding immediate actions to take.

One particular ODS characteristic requires it to handle mixed workloads. On the one hand, it must be able to respond to complex queries from knowledge users, data-mining facilities, and rules engines using online analytical processing (OLAP). The database structures suitable for OLAP are characterized by fat keys that allow rapid searching of the database to respond to complex queries. On the other hand, the ODS must be capable of online transaction processing (OLTP) at an extremely high transaction rate, as it is being fed transactions in real-time from many enterprise systems. The database structures suitable for OLTP are characterized by skinny keys that require a minimum of updating as data is added to the database.

Another particular ODS characteristic is that it is bi-directional. Unlike a data warehouse, which typically only accepts information from enterprise systems, an ODS both accepts information from and delivers information to the other enterprise systems. An example of this characteristic is the act of keeping databases in synchronization. A particular data item, like a customer's address, may be stored in several databases around the enterprise. If one system changes this data item, the ODS acts as a central data repository that informs the other systems of the new data value so they update their databases.

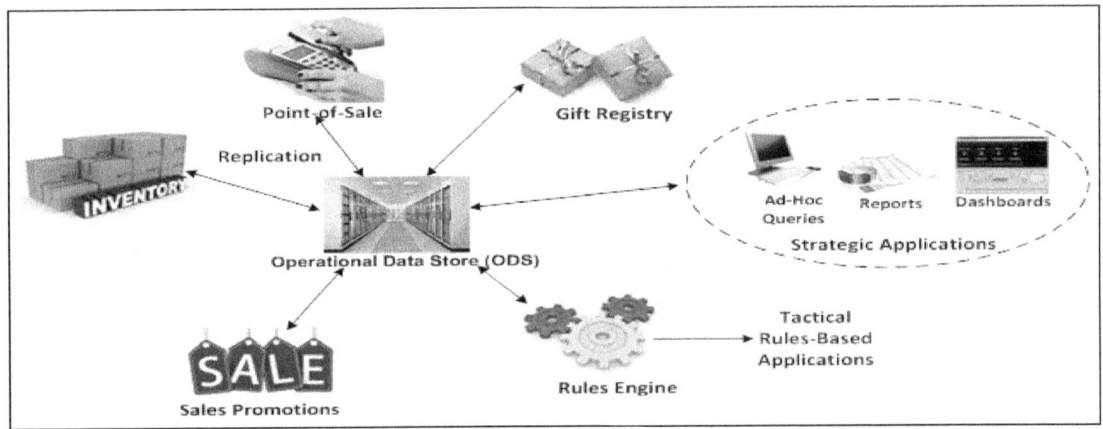

Figure 3.1 - The Operational Data Store

Other examples of outgoing information are the results of the rules engine. If the rules engine decides to recommend a particular immediate action, that action is communicated to the appropriate enterprise system for execution. For instance, if the rules engine for a credit card processor detects suspicious activity, it immediately alerts the authorization system to take appropriate action. In concept, the ODS, which contains all of a corporation's data, could become the database of record, the single version of truth. This action generally does not happen because of regulatory requirements and other considerations and the database of record remains on the existing systems, where it was resident for decades.

3.1 The Evolution of RTBI to ODS

As previously described, the ODS is simply the RTBI system but extended to the enterprise. Though RTBI systems are becoming common today, full-blown ODS systems have yet to make a significant appearance due to the complexity and cost of designing and building such far-reaching systems. It is conceivable that RTBI systems may evolve eventually to ODS systems, as shown in Figure 14. Early dedicated business intelligence systems used online data warehouse or EAI technologies. As these technologies proved their worth, data replication was added to move the technologies to real-time business intelligence systems, thus expanding their reach. Though RTBI systems exist today in many applications, each RTBI system generally supports a single purpose such as fraud detection, instant customer promotions, or just-in-time inventory.

Considerable effort was invested by some companies to consolidate a multitude of RTBI systems into a single ODS supporting enterprise-wide tactical and strategic decision-making or enterprise content management (ECM), The cost and disruption imposed by conversion to an ODS has so far resulted in little progress in expanding RTBI systems to support both tactical and strategic decision-making for any particular corporate function, much less the enterprise.

The advantages of such integration are clear:

- The single ODS supports both tactical and strategic decision making

- The ODS is made highly available through redundancy, such as using an active/active replication system to achieve not only high availability, but also disaster tolerance. Corporate functions are less affected by the failure of one of the other systems. For example, a customer's credit status is checked against the ODS without having to interrogate a credit-authorization system that might be down

- The scope of decision-making is extended too many more areas across the enterprise. For instance, a drop in sales is correlated with increased accounts receivables on store-branded credit cards. Easier credit terms might help to restore sales to their previous level.

- The various corporate IT systems are isolated. No longer do they have to interact with each other through EAI. They each communicate only with the ODS system. Newly added applications do not have to be configured to interface with multiple other applications and only need to interface with the ODS. Also, other systems do not have to be modified to interface with the new system.

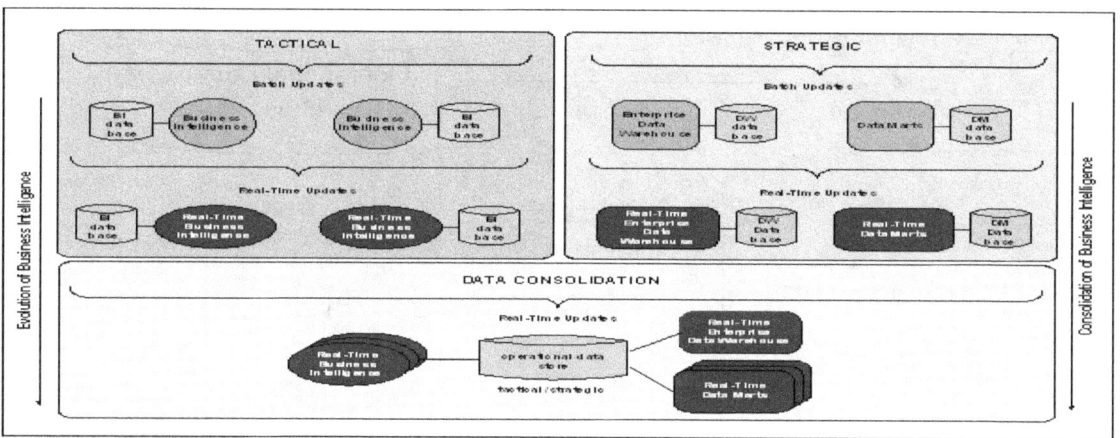

Figure 3.2 - The Evolution of the Operational Data Store

However, so far the obstacles to achieving this goal have been daunting. For instance:

- The ODS is an expensive system in terms of hardware acquisition, software licensing, and development.
- More powerful data-mining engines and rules engines may need to be incorporated.
- The design of the ODS database is much different and needs to support both tactical and strategic queries, yet be very efficient in handling a large volume of updates.
- Fast update processing requires skinny keys in which a minimum of alternate indices must be updated, while efficient query processing requires fat keys providing many access paths to the data.
- The size and depth of the ODS can be quite extensive, as it typically has to store large volumes of historical and archival versions of the data, instead of just the current value of the data (also referred to as managing long data issues).
- Current database products tend to handle structured data well, but corporations also want to leverage their unstructured data (e.g., video, email, voice, text) for business gain. So-called big data collection, indexing, and accessing issues tend to thwart co-

locating all of this information into a single repository, managed by a single DBMS. The volume of data generation overwhelms typical storage capacities. Clearly, more powerful data collection, transmission, storage and retrieval systems are needed.

- Applications may have to be significantly re-architected, which is not only expensive and time consuming, but risky.
- Third-party products that do not readily lend themselves to adapting to an ODS architecture may be involved.
- The conversion of current decision-making processes might not only be difficult but may, in fact, be resisted by the user community.

The bottom line is that today, companies are achieving RTBI by directly integrating their systems using real-time heterogeneous data replication, or they are trickle-feeding data marts in real-time and using these marts to gather information. These warehouses or application networks may or may not turn into an ODS as consolidation occurs. If no warehouse currently exists to act as the stepping stone to an ODS, companies may find it more economical to simply interconnect their systems in a mesh network using existing EAI technologies, instead of following a more planned and fruitful, but expensive path to an ODS.

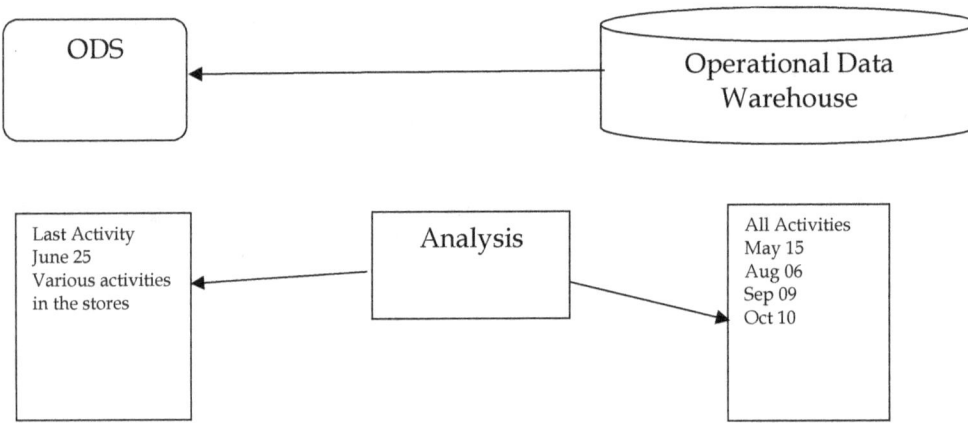

Figure 3.3 - ODS and Data Warehouse

3.2 Evolution of OLAP

Expanding Data Bases and Need for Analysis of corporate data has grown consistently and rapidly during in early the 1980's, businesses worked with data in the megabytes and gigabyte range. Contemporary enterprises are having to manipulate data in the range of terabytes and petabytes. The need for more sophisticated analysis and faster synthesis of better quality information has grown. Business enterprises prosper or fail according to the sophistication and speed of their information systems and their ability to analyze and synthesize information using those systems. The number of individuals within an enterprise who have a need to perform more sophisticated analysis is growing. Business enterprises prosper or fail according to the sophistication and speed of their information systems and their ability to analyze and synthesize information using those systems.

Data in relational systems is also being accessed by a wide variety of non-data processing through the use of many different types of tools and interfaces. The OLAP evolution of analytical processing matured and transformed the data processing needs as growing needs of independently operating reporting systems. These include general purpose query products, spreadsheets, and graphics packages, off the shelf application packages for human resource management, accounting, banking, and other disciplines. The files in which data may be stored and the numerous types of front-end packages that the end users may need. With relational technology, the system complexity of pre-relational systems has been replaced by ease of learning, ease of use, and support for adhoc query and manipulation. Moreover, a relational DBMS includes a more powerful means of preserving the logical integrity of the data than any pre-relational DBMS. At every turn, the relational database management system has become the gateway to enterprise data.

The Evolution of OLAP Of the wide variety of business applications that have been afforded faster, cheaper and better solutions in the relational DBMS world, perhaps none are more dramatic than query/report processing. The OLAP thin layers, client-server, web-based and dashboards proven to be accommodating the needs of data growth, regional presentation of data. Thus empowered, end-users now to a large extent satisfy their own requirements. Not only are they able to experiment with various data formats and aggregations, they are able to improve the information content of their reports. However, as enabling for end-users as these new relational DBMS products and associated tools and interfaces have been, there are still significant limitations to their efficacy. Commercial DBMS products do have boundaries with respect to providing function to support user views of data. OLAP Server Mediating Role of the OLAP Server Hierarchical Spreadsheet, Flat Files, Statistical Package RDBMS and Graphical Interface

The Relational Model dictates relational DBMS system design that provides unprecedented power in storing, updating and retrieving data. The power of any one specific relational DBMS when compared to the power of the Relational Model is dependent on the extent to which the DBMS is faithful to the Relational Model. Most notably lacking has been the ability to consolidate, view, and analyze the data according to multiple dimensions, in ways that make sense to one or more specific enterprise analysts at any given point in time. This requirement is called "multidimensional data analysis." Until recently, the end-user products that had been developed as front-ends to the relational DBMS provided very straightforward simplistic functionality.

The query/report writers and spreadsheets have been extremely limited in the ways in which data (having already been retrieved from the DBMS) can be aggregated, summarized, consolidated, summed, viewed, and analyzed. Most notably lacking has been the ability to consolidate, view, and analyze data according to multiple dimensions, in ways that make sense to one or more specific enterprise analysts at any given point in time. This requirement is called "multidimensional data analysis." Perhaps a better and more generic name for this type of functionality is online analytical processing (OLAP), wherein multidimensional data analysis is but one of its characteristics. The term "OLAP" is defined, its fundamental characteristics including multidimensional data analysis are examined, the business requirement for OLAP is discussed, and the types of users who are likely to benefit most from OLAP are identified.

Relational DBMS were never intended to provide the very powerful functions for data synthesis, analysis, and consolidation that is being defined as multi-dimensional data analysis for creating OLAP functions and Cubes. These types of functions were always intended to be provided by separate, end-user tools that were outside and complementary to the relational DBMS products. The challenge has become one of the functions to get out into market place. Since the end-user has become very comfortable with the interface to the spreadsheet, the most obvious approach would have been to simply add the function to the spreadsheet product. Examination of the collection of functions requiring implementation in order to support OLAP suggests the inclusion of the following:

- Access to the data in the DBMS or access method files
- Definitions of the data and its dimensions required by the user
- The variety of ways and contexts in which the user might wish to view, manipulate, and animate the data model
- Accessibility to these functions via the end-user's customary interface

With the exception of retrieving the data in question, all the other functions listed are really enhancements of the kinds of capabilities one would expect in a spreadsheet product. Unfortunately, the spreadsheet vendors have not shown much interest in supporting online analytical processing to the required degree of robustness as defined by the OLAP evaluation rules. An architectural framework is required within which the function could be made to appear as if it were part of the user analyst's customary spreadsheet product. This framework has become a significant measure of the efficacy of the multidimensional data analysis product itself, and, in fact, represents the first criterion for evaluating OLAP products.

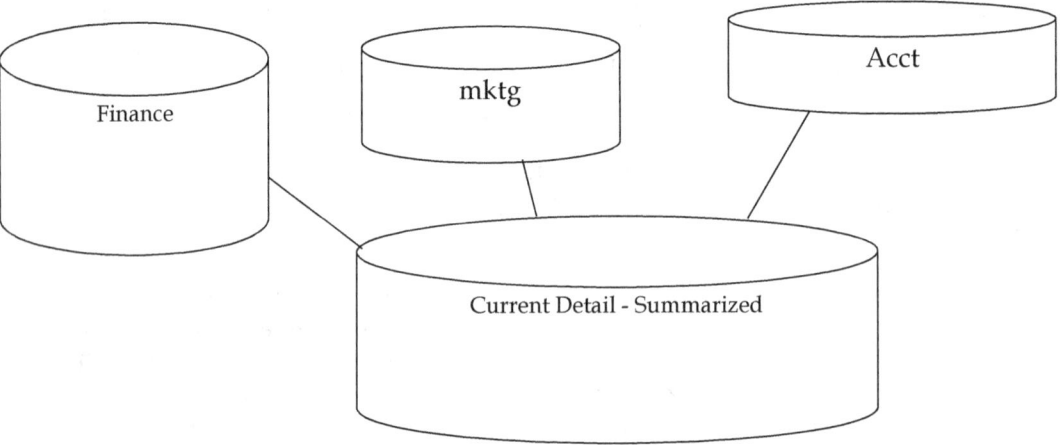

Figure 3.4 – DataMart Multidimensional DBMS (OLAP)

3.3 OLAP CUBE

Cubes are multidimensional data objects. The framework, framework for giving analyst or any application end-user understandable and natural way of reporting. The most detailed unit of the data is a fact. Fact can be a contract, invoice, spending, task, etc. Each fact might have a measure – an attribute that can be measured, such as: price, amount, revenue, duration, tax, discount, etc.

The dimension provides context for facts. Is used to:
- filter queries or reports
- controls scope of aggregation of facts
- used for ordering or sorting
- defines master-detail relationship

Dimension can have multiple hierarchies, for example the date dimension might have year, month and day levels in a hierarchy

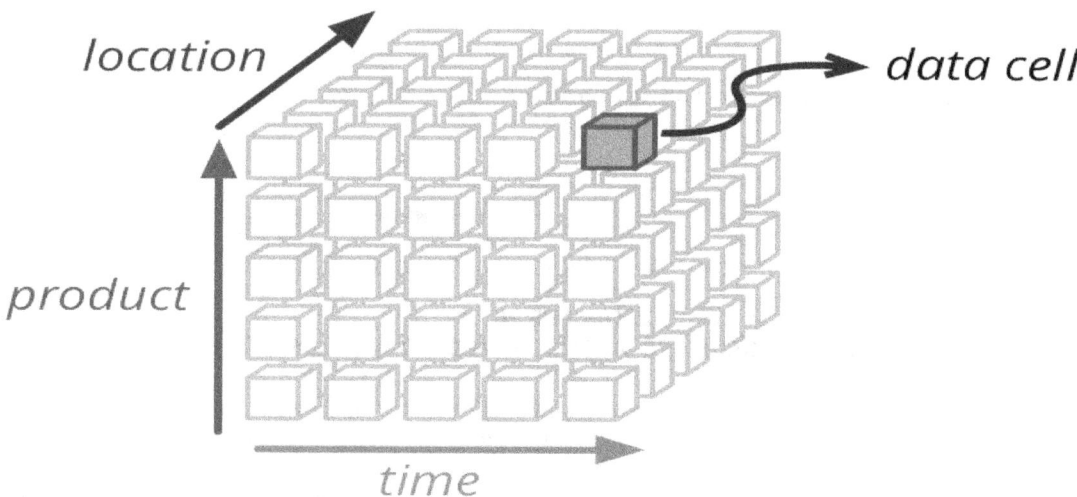

Figure 3.3.1 – A Data Cube

Everything in Cubes happens in an analytical workspace. It contains cubes, maintains connections to the data stores (with cube data), and provides connection to external cubes and more.

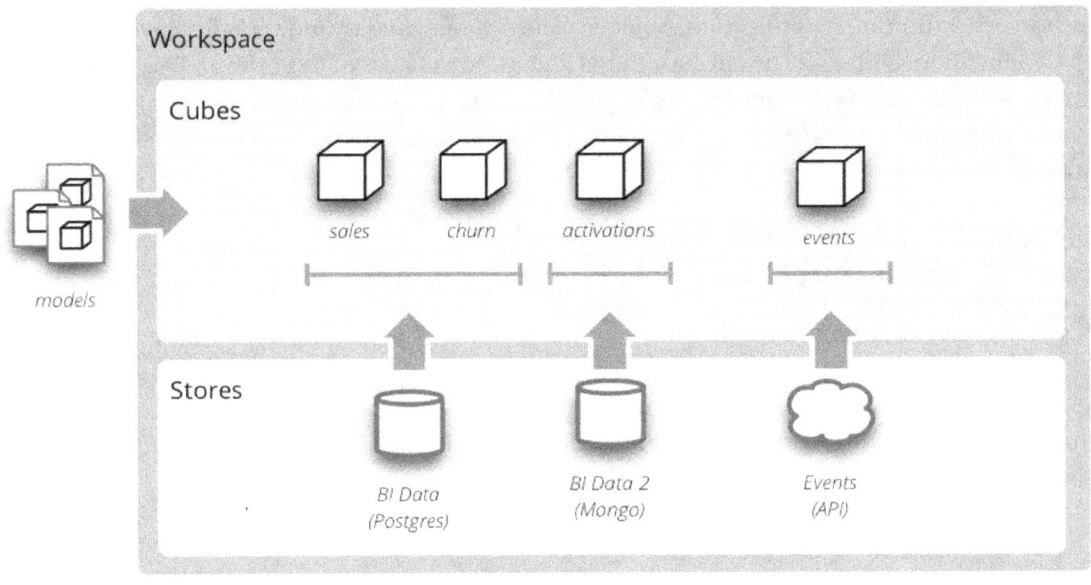

Figure 3.4 - Analytical workspace and its content

CHAPTER 4
ENTERPRISE DATA WAREHOUSING AND CONFIRMED WAY OF DOING BUSINESS

The Enterprise Data Warehouse (EDW) is a service offered by the Data Services department of the information services and technology division. Enterprise Data Warehouse is a centralized warehouse which provides service for the entire enterprise. A data warehouse is by essence a large repository of historical and current transaction data of an organization. EDW consolidates data from multiple sources in support of campus wide decision making and related information needs such as reporting, analysis, and planning. Enterprise data warehouse (EDW), is a system used for reporting and data analysis. Data Warehouse are central repositories of integrated data from one or more disparate sources. They store current and historical data and are used for creating analytical reports for knowledge workers throughout the enterprise. For example the reports could range from annual and quarterly comparisons and trends to detailed daily sales analyses. An Enterprise Data Warehouse is a specialized data warehouse which may have several interpretations. Several terms used in information technology have been used by a so many different vendors, IT workers and marketing ad campaigns that has left many confused about what really the term Enterprise Data Warehouse means and what makes it different from a general data warehouse. Enterprise Data Warehouse has emerged from the convergence of opportunity, capability, infrastructure and need for data which has exponentially increased during the last few years as technology has advanced too fast and Business Enterprises tried to do their best to catch up and be on the top of the industry competition.

In order to give a clear picture of an Enterprise Data Warehouse and how it differs from an ordinary data warehouses, five attributes are being considered. This is not really exclusive they bring people closer to a focused meaning of the Enterprise Data Warehouse from among the many interpretations of the term. These attributes mainly pertain to the overall philosophy as well as the underlying infrastructure of an Enterprise Data Warehouse.

- The first attribute of an Enterprise Data Warehouse is that it should have a single version of truth and that entire goal of the warehouse's design is to come up with a definitive representation of the organization's business data as well as the corresponding rules. Given the number and variety of systems and silos of company data that exist within any business organization, many business warehouses may not qualify as an Enterprise Data Warehouse.

- The second attribute is that an Enterprise Data Warehouse should have multiple subject areas. In order to have a unified version of the truth for an organization, an Enterprise Data Warehouse should contain all subject areas related to the enterprise such as marketing, sale, finance, human resource and others.

- The third attribute is that an Enterprise Data Warehouse should have a normalized design. This may be an arguable attribute as both normalized and de-normalized databases have their own advantages for a data warehouse. In fact, may data warehouse designers have used de-normalized models such as star or snowflake schemas for implementing datamarts . But many also go for normalized databases for an Enterprise Data Warehouse in the consideration of flexibility first and performance second.

- The fourth attribute is that an Enterprise Data Warehouse should be implemented as a Mission-Critical Environment. The entire underlying infrastructure should be able to handle any unforeseen critical conditions because failure in the data warehouse means stoppage of the business operation and loss of income and revenue. An Enterprise Data Warehouse should have high availability features such as online parameter or database structural changes, business continuance such as failover and disaster recovery features and security features. The EDW should also possess the capability of handling the data re-consolation process without affecting the business.

- The fifth and finally attribute of an Enterprise Data Warehouse should be scalable across several dimensions. It should expect that a company's main objective is to grow and that the warehouse should be able to handle the growth of data as well as the growing complexities of processes which will come together with the evolution of the business enterprise. Because of the fast evolution of information technology, many business rules have been changed or broken to make way for rules which are data driven. Processes may fluctuate from simple to complex and data may shrink or grow in the constantly changing enterprise environment. Hence, a real Enterprise Data Warehouse should scale to the changes.

4.1 Top-down Design

In the top-down design approach the, data warehouse is built first. The data marts are then created from the data warehouse. The top-down approach is designed using a normalized enterprise data model. "Atomic" data, that is, data at the lowest level of detail, are stored in the data warehouse. Dimensional data marts containing data needed for specific business processes or specific departments are created from the data warehouse.

Advantages of top-down design:

- Provides consistent dimensional views of data across data marts, as all data marts are loaded from the data warehouse. This approach is robust against business changes. Creating a new data mart from the data warehouse is very easy.

Disadvantages of top-down design:

- This methodology is inflexible to changing departmental needs during implementation phase. It represents a very large project and the cost of implementing the project is significant.

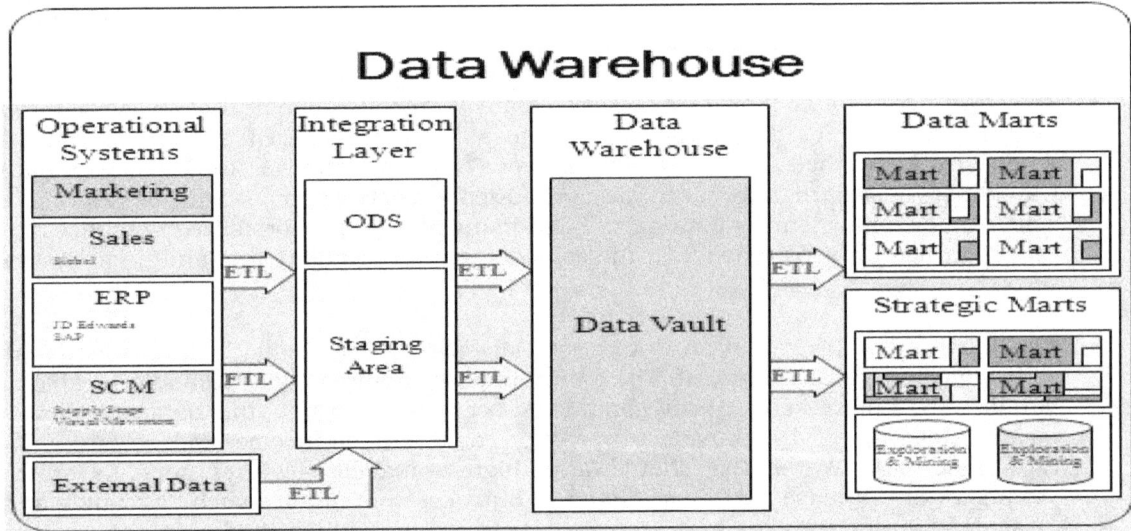

Figure 4.1 – Enterprise Data Warehousing an Example with a Top-down Approach and Atomic Data Warehouse

4.2 Bottom-Up Design:

In the bottom-up design approach, the data marts are created first to provide reporting capability. A data mart addresses a single business area such as sales, Finance etc. These data marts are then integrated to build a complete data warehouse. The integration of data marts is implemented using data warehouse bus architecture. In the bus architecture, a dimension is shared between facts in two or more data marts. These dimensions are called conformed dimensions. These conformed dimensions are integrated from data marts and then data warehouse is built. In the bottom-up approach, data marts are first created to provide reporting and analytical capabilities for specific business processes. These data marts can then be integrated to create a comprehensive data warehouse. The data warehouse bus architecture is primarily an implementation of "the bus", a collection of conformed dimensions and conformed facts, which are dimensions that are shared (in a specific way) between facts in two or more data marts.

Advantages of bottom-up design:

This model contains consistent data marts and these data marts can be delivered quickly. As the data marts are created first, reports can be generated quickly. The data warehouse can be extended easily to accommodate new business units. It is just creating new data marts and then integrating with other data marts.

Disadvantages of bottom-up design:

The positions of the data warehouse and the data marts are reversed in the bottom-up approach design.

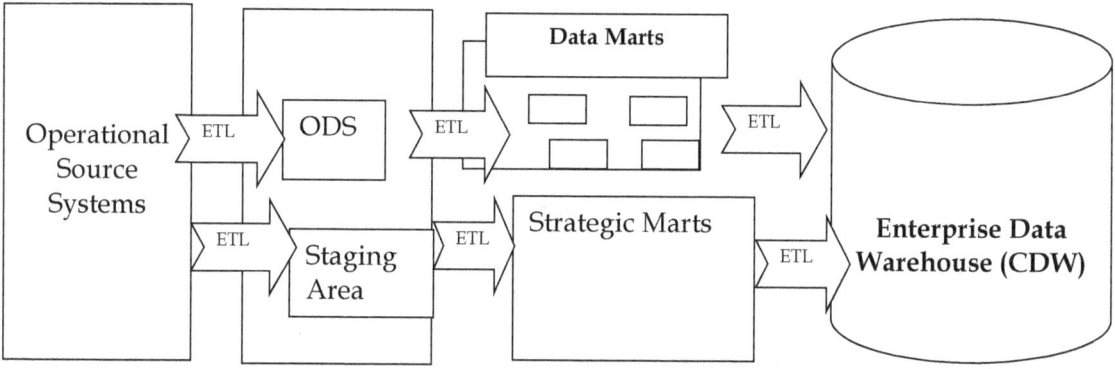

Figure 4.2 – Enterprise Data Warehousing an Example with a Bottom-Up Approach and CDW

4.3 Types of Systems

4.3.1 Data mart

A data mart is a simple form of a data warehouse that is focused on a single subject (or functional area), such as sales, finance or marketing. Data marts are often built and controlled by a single department within an organization. Given their single-subject focus, data marts usually draw data from only a few sources. The sources could be internal operational systems, a central data warehouse, or external data. De-normalization is the norm for data modeling techniques in this system.

4.3.2 Online analytical processing (OLAP)

Is characterized by a relatively low volume of transactions. Queries are often very complex and involve aggregations. For OLAP systems, response time is an effectiveness measure. OLAP applications are widely used by Data Mining techniques. OLAP databases store aggregated, historical data in multi-dimensional schemas (usually star schemas). OLAP systems typically have data latency of a few hours, as opposed to data marts, where latency is expected to be closer to one day.

4.3.3 Online Transaction Processing (OLTP)

OLTP characterized by a large number of short on-line transactions (INSERT, UPDATE, and DELETE). OLTP systems emphasize very fast query processing and maintaining data integrity in multi-access environments. For OLTP systems, effectiveness is measured by the number of transactions per second. OLTP databases contain detailed and current data. The schema used to store transactional databases is the entity model (usually 3NF). Normalization is the norm for data modeling techniques in this system.

4.3.4 Predictive analysis

Predictive analysis is about finding and quantifying hidden patterns in the data using complex mathematical models that can be used to predict future outcomes. Predictive analysis is different from OLAP in that OLAP focuses on historical data analysis and is reactive in nature, while predictive analysis focuses on the future. These systems are also used for CRM (Customer Relationship Management). A data warehouse maintains a copy of information from the source transaction systems. This architectural complexity provides the opportunity to:

- Congregate data from multiple sources into a single database so a single query engine can be used to present data.
- Mitigate the problem of database isolation level lock contention in transaction processing systems caused by attempts to run large, long running, analysis queries in transaction processing databases.
- Maintain data history, even if the source transaction systems do not.
- Integrate data from multiple source systems, enabling a central view across the enterprise. This benefit is always valuable, but particularly so when the organization has grown by merger.
- Improve data quality, by providing consistent codes and descriptions, flagging or even fixing bad data.

- Present the organization's information consistently.
- Provide a single common data model for all data of interest regardless of the data's source.
- Restructure the data so that it makes sense to the business users.
- Restructure the data so that it delivers excellent query performance, even for complex analytic queries, without impacting the operational systems.
- Add value to operational business applications, notably customer relationship management (CRM) systems.
- Make decision–support queries easier to write.

4.3.5 Data Warehouses versus Operational Systems

Operational systems are optimized for preservation of data integrity and speed of recording of business transactions through use of database normalization and an entity-relationship model. Operational system designers generally follow the Codd rules of database normalization in order to ensure data integrity. Codd defined five increasingly stringent rules of normalization. Fully normalized database designs (that is, those satisfying all five Codd rules) often result in information from a business transaction being stored in dozens to hundreds of tables. Relational databases are efficient at managing the relationships between these tables. The databases have very fast insert/update performance because only a small amount of data in those tables is affected each time a transaction is processed. Finally, in order to improve performance, older data are usually periodically purged from operational systems.

Data warehouses are optimized for analytic access patterns. Analytic access patterns generally involve selecting specific fields and rarely if ever 'select *' as is more common in operational databases. Because of these differences in access patterns, operational databases (loosely, OLTP) benefit from the use of a row-oriented DBMS whereas analytics databases (loosely, OLAP) benefit from the use of a column-oriented DBMS. Unlike operational systems which maintain a snapshot of the business, data warehouses generally maintain an infinite history which is implemented through ETL processes that periodically migrate data from the operational systems over to the data warehouse.

4.3.6 Offline operational data warehouse

Data warehouses in this stage of evolution are updated on a regular time cycle (usually daily, weekly or monthly) from the operational systems and the data is stored in an integrated reporting-oriented data.

Offline data warehouse

Data warehouses at this stage are updated from data in the operational systems on a regular basis and the data warehouse data are stored in a data structure designed to facilitate reporting.

On time data warehouse

Online Integrated Data Warehousing represent the real time Data warehouses stage data in the warehouse is updated for every transaction performed on the source data

4.3.7 Integrated data warehouse

These data warehouses assemble data from different areas of business, so users can look up the information they need across other systems.

4.4 Star Schema

In computing, the Star Schema is the simplest style of data mart schema. The star schema consists of one or more fact tables referencing any number of dimension tables. The star schema is an important special case of the snowflake schema, and is more effective for handling simpler queries. The star schema gets its name from the physical model's resemblance to a star with a fact table at its center and the dimension tables surrounding it representing the star's points.

The star schema separates business process data into facts, which hold the measurable, quantitative data about a business, and dimensions which are descriptive attributes related to fact data. Examples of fact data include sales price, sale quantity, and time, distance, speed, and weight measurements. Related dimension attribute examples include product models, product colors, product sizes, geographic locations, and salesperson names. A star schema that has many dimensions is sometimes called a centipede schema. Having dimensions of only a few attributes, while simpler to maintain, results in queries with many table joins and makes the star schema less easy to use.

4.4.1 Fact Tables

Fact tables record measurements or metrics for a specific event. Fact tables generally consist of numeric values, and foreign keys to dimensional data where descriptive information is kept. Fact tables are designed to a low level of uniform detail (referred to as "granularity" or "grain"), meaning facts can record events at a very atomic level. This can result in the accumulation of a large number of records in a fact table over time. Fact tables are defined as one of three types:

- Transaction fact tables record facts about a specific event (e.g., sales events)
- Snapshot fact tables record facts at a given point in time (e.g., account details at month end)
- Accumulating snapshot tables record aggregate facts at a given point in time (e.g., total month-to- date sales for a product)
- Fact tables are generally assigned a surrogate key to ensure each row can be uniquely identified.

4.4.2 Dimension tables

Dimension tables usually have a relatively small number of records compared to fact tables, but each record may have a very large number of attributes to describe the fact data.

Dimensions can define a wide variety of characteristics, but some of the most common attributes defined by dimension tables include:

- Time dimension tables describe time at the lowest level of time granularity for which events are recorded in the star schema
- Geography dimension tables describe location data, such as country, state, or city
- Product dimension tables describe products
- Employee dimension tables describe employees, such as sales people
- Range dimension tables describe ranges of time, dollar values, or other measurable quantities to simplify reporting
- Dimension tables are generally assigned a surrogate primary key, usually a single-column integer data type, mapped to the combination of dimension attributes that form the natural key.

4.4.3 Benefits

Star schemas are de-normalized, meaning the normal rules of normalization applied to transactional relational databases are relaxed during star schema design and implementation. The benefits of star schema de-normalization are:

- Simpler queries - star schema join logic is generally simpler than the join logic required to retrieve data from a highly normalized transactional schemas.
- Simplified business reporting logic - when compared to highly normalized schemas, the star schema simplifies common business reporting logic, such as period-over-period and as-of reporting.
- Query performance gains - star schemas can provide performance enhancements for read-only reporting applications when compared to highly normalized schemas.
- Fast aggregations - the simpler queries against a star schema can result in improved performance for aggregation operations.
- Feeding cubes - star schemas are used by all OLAP systems to build proprietary OLAP cubes efficiently; in fact, most major OLAP systems provide a ROLAP mode of operation which can use a star schema directly as a source without building a proprietary cube structure.

4.4.4 Dis–Advantages

The main disadvantage of the star schema is that data integrity is not enforced as well as it is in a highly normalized database. One-off inserts and updates can result in data anomalies which normalized schemas are designed to avoid. Generally speaking, star schemas are loaded in a highly controlled fashion via batch processing or near-real time "trickle feeds", to compensate for the lack of protection afforded by normalization.

Star schema is also not as flexible in terms of analytical needs as a normalized data model. Normalized models allow any kind of analytical queries to be executed as long as they follow the business logic defined in the model. Star schemas tend to be more purpose-built for a particular view of the data, thus not really allowing more complex analytics. Star schemas don't support many-to-many relationships between business entities - at least not very naturally. Typically these relationships are simplified in star schema to conform the simple dimensional model.

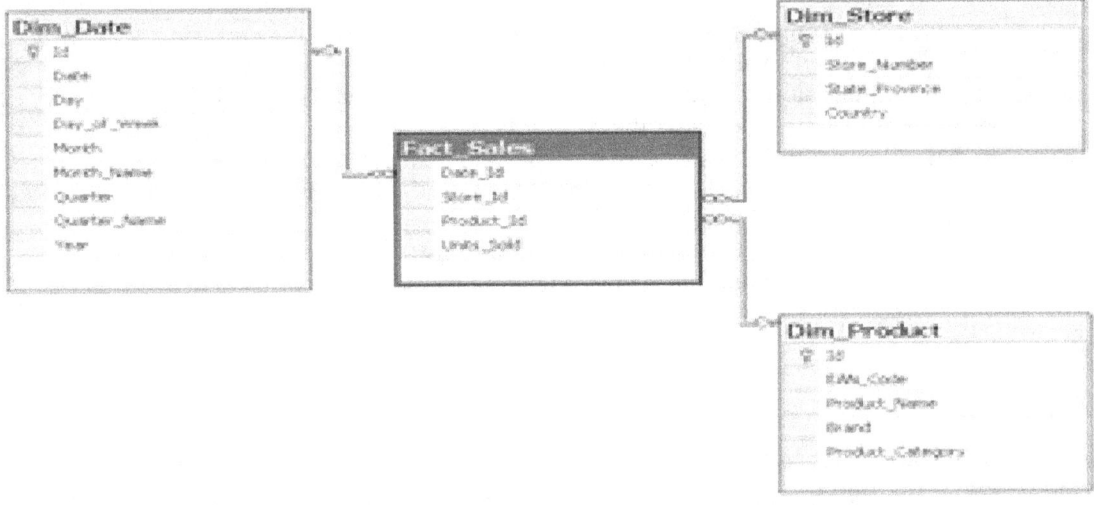

Figure 4.4.1 - Star Schema

Consider a database of sales, perhaps from a store chain, classified by date, store and product. The image of the schema to the right is a star schema version of the sample schema provided in the snowflake schema article. Fact_Sales is the fact table and there are three dimension tables Dim_Date, Dim_Store and Dim_Product. Each dimension table has a primary key on its Id column, relating to one of the columns (viewed as rows in the example schema) of the Fact_Sales table's three-column (compound) primary key (Date_Id, Store_Id, Product_Id). The non-primary key Units_Sold column of the fact table in this example represents a measure or metric that can be used in calculations and analysis. The non-primary key columns of the dimension tables represent additional attributes of the dimensions (such as the Year of the Dim_Date dimension).

For example, the following query answers how many TV sets have been sold, for each brand and country for period in 1997:

SELECT
     P.Brand,
     S.Country as Countries,
     SUM (F.Units_Sold)

FROM Fact_Sales F
INNER JOIN Dim_Date D   ON (F.Date_Id = D.Id)
INNER JOIN Dim_Store S   ON (F.Store_Id = S.Id)
INNER JOIN Dim_Product P ON (F.Product_Id = P.Id)
WHERE D.Year = 1997 AND P.Product_Category = 'tv'
GROUP BY
     P.Brand,
     S.Country

4.5 Snowflake Schema

Snowflake schema is a logical arrangement of tables in a multidimensional database such that the entity relationship diagram resembles a snowflake shape. The snowflake schema is represented by centralized fact tables which are connected to multiple dimensions. "Snowflaking" is a method of normalizing the dimension tables in a star schema. When it is completely normalized along all the dimension tables, the resultant structure resembles a snowflake with the fact table in the middle. The principle behind snow flaking is normalization of the dimension tables by removing low cardinality attributes and forming separate tables.

The snowflake schema is similar to the star schema. However, in the snowflake schema, dimensions are normalized into multiple related tables, whereas the star schema's dimensions are de-normalized with each dimension represented by a single table. A complex snowflake shape emerges when the dimensions of a snowflake schema are elaborate, having multiple levels of relationships, and the child tables have multiple parent tables ("forks in the road").

4.5.1 Common Uses

Star and snowflake schemas are most commonly found in dimensional data warehouses and data marts where speed of data retrieval is more important than the efficiency of data manipulations. As such, the tables in these schemas are not normalized much, and are frequently designed at a level of normalization short of third normal form.

Deciding whether to deploy a star schema or a snowflake schema should involve considering the relative strengths of the database platform in question and the query tool to be employed. Star schemas should be favored with query tools that largely expose users to the underlying table structures, and in environments where most queries are simpler in nature. Snowflake schemas are often better with more sophisticated query tools that create a layer of abstraction between the users and raw table structures for environments having numerous queries with complex criteria.

4.5.2 Data Normalization Storage

Normalization splits up data to avoid redundancy (duplication) by moving commonly repeating groups of data into new tables. Normalization therefore tends to increase the number of tables that need to be joined in order to perform a given query, but reduces the space required to hold the data and the number of places where it needs to be updated if the data changes. From a space storage point of view, the dimensional tables are typically small compared to the fact tables. This often removes the storage space benefit of snow flaking the dimension tables, as compared with a star schema. Some database developers compromise by creating an underlying snowflake schema with views built on top of it that perform many of the necessary joins to simulate a star schema. This provides the storage benefits achieved through the normalization of dimensions with the ease of querying that the star schema provides. The tradeoff is that requiring the server to perform the underlying joins automatically can result in a performance hit when querying as well as extra joins to tables that may not be necessary to fulfill certain queries. The snowflake schema is in the same family as the star schema logical model. In fact, the star schema is considered a special case of the snowflake schema. The snowflake schema provides some advantages over the star schema in certain situations, including:

4.5.3 Benefits

- Some OLAP multidimensional database modeling tools are optimized for snowflake schemas.
- Normalizing attributes results in storage savings, the tradeoff being additional complexity in source query joins.

4.5.6 Dis-Advantages

The primary disadvantage of the snowflake schema is that the additional levels of attribute normalization adds complexity to source query joins, when compared to the star schema. Snowflake schemas, in contrast to flat single table dimensions, have been heavily criticized. Their goal is assumed to be an efficient and compact storage of normalized data but this is at the significant cost of poor performance when browsing the joins required in this dimension. This disadvantage may have reduced in the years since it was first recognized, owing to better query performance within the browsing tools. When compared to a highly normalized transactional schema, the snowflake schema's denormalization removes the data integrity assurances provided by normalized schemas. Data loads into the snowflake schema must be highly controlled and managed to avoid update and insert anomalies.

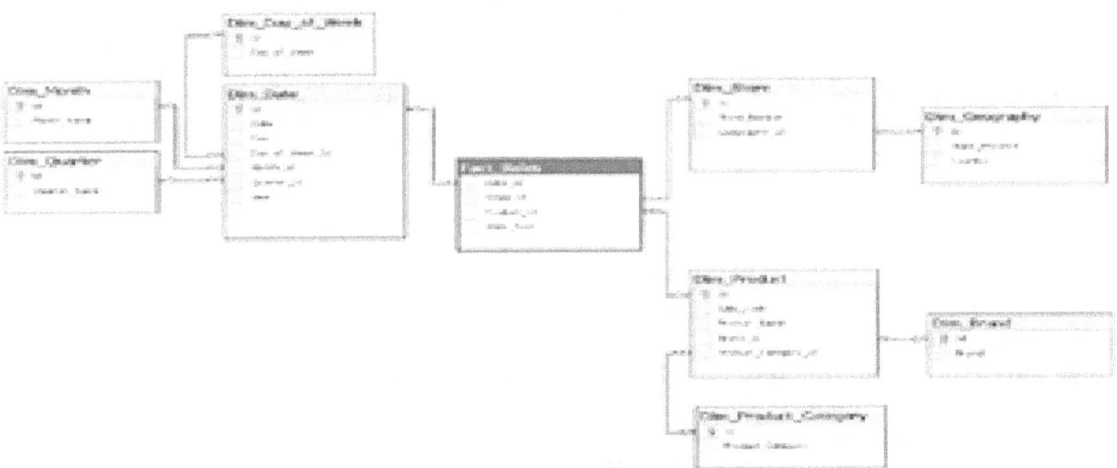

Figure 4.5.1 - Snowflake Schema Dimensional Data Model

4.5.6 Snowflake schema used by Example Query:

The example schema shown to the right is a snowflaked version of the star schema example provided in the star schema article. The following example query is the snowflake schema equivalent of the star schema example code which returns the total number of units sold by brand and by country for 1997. Notice that the snowflake schema query requires many more joins than the star schema version in order to fulfill even a simple query. The benefit of using the snowflake schema in this example is that the storage requirements are lower since the snowflake schema eliminates many duplicate values from the dimensions themselves.

```
SELECT
        B.Brand,
        G.Country,
        SUM(F.Units_Sold)
FROM Fact_Sales F
INNER JOIN Dim_Date D          ON F.Date_Id = D.Id
INNER JOIN Dim_Store S         ON F.Store_Id = S.Id
INNER JOIN Dim_Geography G      ON S.Geography_Id = G.Id
INNER JOIN Dim_Product P       ON F.Product_Id = P.Id
INNER JOIN Dim_Brand B         ON P.Brand_Id = B.Id
INNER JOIN Dim_Product_Category C ON P.Product_Category_Id = C.Id
WHERE
        D.Year = 1997 AND
        C.Product_Category = 'tv'
GROUP BY
        B.Brand,
        G.Country
```

4.6 Slowly Changing Dimensions

Dimensions in data management and data warehousing contain relatively static data about such entities as geographical locations, customers, or products. Data captured by Slowly Changing Dimensions (SCDs) change slowly but unpredictably, rather than according to a regular schedule. Some scenarios can cause Referential integrity problems.

For example, a database may contain a fact table that stores sales records. This fact table would be linked to dimensions by means of foreign keys. One of these dimensions may contain data about the company's salespeople: e.g., the regional offices in which they work. However, the salespeople are sometimes transferred from one regional office to another. For historical sales reporting purposes it may be necessary to keep a record of the fact that a particular sales person had been assigned to a particular regional office at an earlier date, whereas that sales person is presently assigned to a different regional office. Dealing with these issues involves SCD management methodologies referred to as Type 0 through 6. Type 6 SCDs are also sometimes called Hybrid SCDs.

4.6.1 Type 0

The Type 0 method is passive. It manages dimensional changes and no action is performed. Values remain as they were at the time the dimension record was first inserted. In certain circumstances history is preserved with a Type 0. High order types are employed to guarantee the preservation of history whereas Type 0 provides the least or no control. This is rarely used.

4.6.2 Type 1

This methodology overwrites old with new data, and therefore does not track historical data.

Example of a supplier table:

Supplier_Key	Supplier_Code	Supplier_Name	Supplier_State
123	ABC	Acme Supply Co	CA

In the above example, Supplier_Code is the natural key and Supplier_Key is a surrogate key. Technically, the surrogate key is not necessary, since the row will be unique by the natural key (Supplier_Code). However, to optimize performance on joins use integer rather than character keys (unless the number of bytes in the character key is less than the number of bytes in the integer key).

If the supplier relocates the headquarters to Illinois the record would be overwritten:

Supplier_Key	Supplier_Code	Supplier_Name	Supplier_State
123	ABC	Acme Supply Co	IL

The disadvantage of the Type 1 method is that there is no history in the data warehouse. It has the advantage however that it's easy to maintain. If you have calculated an aggregate table summarizing facts by state, it will need to be recalculated when the Supplier_State is changed.

4.6.3 Type 2

This method tracks historical data by creating multiple records for a given natural key in the dimensional tables with separate surrogate keys and/or different version numbers. Unlimited history is preserved for each insert.

For example, if the supplier relocates to Illinois the version numbers will be incremented sequentially:

Supplier_Key	Supplier_Code	Supplier_Name	Supplier_State	Version.
123	ABC	Acme Supply Co	CA	0
124	ABC	Acme Supply Co	IL	1

Another method is to add 'effective date' columns.

Supplier_Key	Supplier_Code	Supplier_Name	Supplier_State	Start_Date	End_Date
123	ABC	Acme Supply Co	CA	01-Jan-2000	21-Dec-2004

The null End_Date in row two indicates the current tuple version. In some cases, a standardized surrogate high date (e.g. 9999-12-31) may be used as an end date, so that the field can be included in an index, and so that null-value substitution is not required when querying. Transactions that reference a particular surrogate key (Supplier_Key) are then permanently bound to the time slices defined by that row of the slowly changing dimension table. An aggregate table summarizing facts by state continues to reflect the historical state, i.e. the state the supplier was in at the time of the transaction; no update is needed. To reference natural key is necessary remove the unique making Referential integrity by DBMS impossible.

If there are retrospective changes made to the contents of the dimension, or if new attributes are added to the dimension (for example a Sales_Rep column) which have different effective dates from those already defined, then this can result in the existing transactions needing to be updated to reflect the new situation. This can be an expensive database operation, so Type 2 SCDs are not a good choice if the dimensional model is subject to change

4.6.4 Type 3

This method tracks changes using separate columns and preserves limited history. The Type 3 preserves limited history as it is limited to the number of columns designated for storing historical data. The original table structure in Type 1 and Type 2 is the same but Type 3 adds additional columns. In the following example, an additional column has been added to the table to record the supplier's original state - only the previous history is stored.

Supplier_Key	Supplier_Code	Supplier_Name	Original_Supplier_State	Effective_Date	Current_Supplier_State
123	ABC	Acme Supply Co	CA	22-Dec-2004	IL

This record contains a column for the original state and current state—cannot track the changes if the supplier relocates a second time. One variation of this is to create the field Previous_Supplier_State instead of Original_Supplier_State which would track only the most recent historical change.

4.6.5 Type 4

The Type 4 method is usually referred to as using "history tables", where one table keeps the current data, and an additional table is used to keep a record of some or all changes. Both the surrogate keys are referenced in the Fact table to enhance query performance. For the above example the original table name is Supplier and the history table is Supplier_History.

Supplier

Supplier_key	Supplier_Code	Supplier_Name	Supplier_State
123	ABC	Acme & Johnson Supply Co	IL

Supplier_History

Supplier_key	Supplier_Code	Supplier_Name	Supplier_State	Create_Date
123	ABC	Acme Supply Co	CA	14-June-2003
123	ABC	Acme & Johnson Supply Co	IL	22-Dec-2004

This method resembles how database audit tables and **change data capture** techniques function.

4.6.6 Type 6/Hybrid

The Type 6 method combines the approaches of types 1, 2 and 3 (1 + 2 + 3 = 6). One possible explanation of the origin of the term was that it was coined by Ralph Kimball during a conversation with Stephen Pace from Kalido. Ralph Kimball calls this method "Unpredictable Changes with Single-Version Overlay in The Data Warehouse Toolkit.

The Supplier table starts out with one record for our example supplier:

Supplier_Key	Supplier_Code	Supplier_Name	Current_State	Historical_State	Start_Date	End_Date	Current_Flag
123	ABC	Acme Supply Co	CA	CA	01-Jan-2000	31-Dec-9999	Y

The Current_State and the Historical_State are the same. The Current_Flag attribute indicates that this is the current or most recent record for this supplier. When Acme Supply Company moves to Illinois, we add a new record, as in Type 2 processing: We overwrite the Current_State information in the first record (Supplier_Key = 123) with the new information, as in Type 1 processing. We create a new record to track the changes, as in Type 2 processing. And we store the history in a second State column (Historical_State), which incorporates Type 3 processing.

Supplier_Key	Supplier_Code	Supplier_Name	Current_State	Historical_State	Start_Date	End_Date	Current_Flag
123	ABC	Acme Supply Co	IL	CA	01-Jan-2000	21-Dec-2004	N
124	ABC	Acme Supply Co	IL	IL	22-Dec-2004	31-Dec-9999	Y

For example if the supplier were to relocate again, we would add another record to the Supplier dimension, and we would overwrite the contents of the Current_State column:

Supplier_Key	Supplier_Code	Supplier_Name	Current_State	Historical_State	Start_Date	End_Date	Current_Flag
123	ABC	Acme Supply Co	NY	CA	01-Jan-2000	21-Dec-2004	N
124	ABC	Acme Supply Co	NY	IL	22-Dec-2004	03-Feb-2008	N
125	ABC	Acme Supply Co	NY	NY	04-Feb-2008	31-Dec-99	Y

Note that, for the current record (Current_Flag = 'Y'), the Current_State and the Historical_State are always the same.

4.6.6 Type2/Type6 Fact Implementation

In many Type 2 and Type 6 SCD implementations, the **surrogate key** from the dimension is put into the fact table in place of the **natural key** when the fact data is loaded into the data repository.[1] The surrogate key is selected for a given fact record based on its effective date and the Start_Date and End_Date from the dimension table. This allows the fact data to be easily joined to the correct dimension data for the corresponding effective date.

Type 2 surrogate key with Type 3 attribute

Here is the Supplier table as we created it above using Type 6 Hybrid methodology:

Sup plier _Key	Supplier_ Code	Supplier_ Name	Current_ State	Historical_ State	Start_ Date	End_ Date	Current_Flag
123	ABC	Acme Supply Co	NY	CA	01-Jan-2000	21-Dec-2004	N
124	ABC	Acme Supply Co	NY	IL	22-Dec-2004	03-Feb-2008	N
125	ABC	Acme Supply Co	NY	NY	04-Feb-2008	31-Dec-9999	Y

Once the Delivery table contains the correct Supplier_Key, it can easily be joined to the Supplier table using that key. The following SQL retrieves, for each fact record, the current supplier state and the state the supplier was located in at the time of the delivery:

```
SELECT delivery.delivery_cost, supplier.supplier_name,
supplier.historical_state,supplier.current_state
FROM delivery
INNER JOIN supplier
ON delivery.supplier_key = supplier.supplier_key
```

4.6.7 Pure Type 6 implementation

Having a Type 2 surrogate key for each time slice can cause problems if the dimension is subject to change. A pure Type 6 implementation does not use this, but uses a Surrogate Key for each master data item (e.g. each unique supplier has a single surrogate key). This avoids any changes in the master data having an impact on the existing transaction data. It also allows more options when querying the transactions. Here is the Supplier table using the pure Type 6 methodology:

Supplier_Key	Supplier_Code	Supplier_Name	Supplier_State	Start_Date	End_Date
456	ABC	Acme Supply Co	CA	01-Jan-2000	21-Dec-2004
456	ABC	Acme Supply Co	IL	22-Dec-2004	03-Feb-2008
456	ABC	Acme Supply Co	NY	04-Feb-2008	31-Dec-9999

The following example shows how the query must be extended to ensure a single supplier record is retrieved for each transaction.

SELECT supplier.supplier_code,supplier.supplier_state FROM supplier
INNER JOIN delivery ON supplier.supplier_key = delivery.supplier_key
AND delivery.delivery_date BETWEEN supplier.start_date AND supplier.end_date

A fact record with an effective date (Delivery_Date) of August 9, 2001 will be linked to Supplier_Code of ABC, with a Supplier_State of 'CA'. A fact record with an effective date of October 11, 2007 will also be linked to the same Supplier_Code ABC, but with a Supplier_State of 'IL'.

Whilst more complex, there are a number of advantages of this approach, including: Referential integrity by DBMS is now possible. If there is more than one date on the fact (e.g. Order Date, Delivery Date, Invoice Payment Date) you can choose which date to use for a query.

You can do "as at now", "as at transaction time" or "as at a point in time" queries by changing the date filter logic. You don't need to reprocess the Fact table if there is a change in the dimension table (e.g. adding additional fields retrospectively which change the time slices, or if you make a mistake in the dates on the dimension table you can correct them easily).You can introduce bi-temporal dates in the dimension table.You can join the fact to the multiple versions of the dimension table to allow reporting of the same information with different effective dates, in the same query. The following example shows how a specific date such as '2012-01-01 00:00:00' (which could be the current datetime) can be used.

SELECT supplier.supplier_code, supplier.supplier_state FROM supplier INNER JOIN delivery ON supplier.supplier_key = delivery.supplier_key AND '2012-01-01 00:00:00' BETWEEN supplier.start_date AND supplier.end_date

Both surrogate and natural key

An alternative implementation is to place both the surrogate key and the natural key into the fact table. This allows the user to select the appropriate dimension records based on: the primary effective date on the fact record (above), the most recent or current information, any other date associated with the fact record. This method allows more flexible links to the dimension, even if you have used the Type 2 approach instead of Type 6.

Here is the Supplier table as we might have created it using Type 2 methodology:

Supplier_ Key	Supplier_ Code	Supplier_Na me	Supplier_St ate	Start_D ate	End_Da te	Current_Flag
123	ABC	Acme Supply Co	CA	01-Jan-2000	21-Dec-2004	N
124	ABC	Acme Supply Co	IL	22-Dec-2004	03-Feb-2008	N
125	ABC	Acme Supply Co	NY	04-Feb-2008	31-Dec-9999	Y

The following SQL retrieves the most current Supplier_Name and Supplier_State for each fact record:

SELECT delivery.delivery_cost, supplier.supplier_name,supplier.supplier_state,FROM delivery INNER JOIN supplier ON delivery.supplier_code = supplier.supplier_code WHERE supplier.current_flag = 'Y'

If there are multiple dates on the fact record, the fact can be joined to the dimension using another date instead of the primary effective date. For instance, the Delivery table might have a primary effective date of Delivery_Date, but might also have an Order_Date associated with each record.

The following SQL retrieves the correct Supplier_Name and Supplier_State for each fact record based on the Order_Date:

SELECT delivery.delivery_cost, supplier.supplier_name,supplier.supplier_state FROM delivery INNER JOIN supplier ON delivery.supplier_code = supplier.supplier_code AND delivery.order_date BETWEEN supplier.start_date AND supplier.end_date

4.6.8 Combining Types

Different SCD Types can be applied to different columns of a table. For example, we can apply Type 1 to the Supplier_Name column and Type 2 to the Supplier_State column of the same table.

4.7 Hybrid design

Data warehouses (DW) often resemble the hub and spokes architecture. Legacy systems feeding the warehouse often include customer relationship management and enterprise resource planning, generating large amounts of data. To consolidate these various data models, and facilitate the extract transform load process, date warehouses often make use of an operational data store, the information from which is parsed into the actual DW. To reduce data redundancy, larger systems often store the data in a normalized way. Data marts for specific reports can then be built on top of the DW.

The DW database in a hybrid solution is kept on third normal form to eliminate data redundancy. A normal relational database, however, is not efficient for business intelligence reports where dimensional modelling is prevalent. Small data marts can shop for data from the consolidated warehouse and use the filtered, specific data for the fact tables and dimensions required. The DW provides a single source of information from which the data marts can read, providing a wide range of business information. The hybrid architecture allows a DW to be replaced with a master data management solution where operational, not static information could reside.

The components follow hub and spokes architecture. This modelling style is a hybrid design, consisting of the best practices from both third normal form and star schema. The Data Vault model is not a true third normal form, and breaks some of its rules, but it is a top-down architecture with a bottom up design. The Data Vault model is geared to be strictly a data warehouse. It is not geared to be end-user accessible, which when built, still requires the use of a data mart or star schema based release area for business purposes.

4.8 Confirmed way of doing the Business

In data warehousing, a conformed dimension is a dimension that has the same meaning to every fact with which it relates. Conformed dimensions allow facts and measures to be categorized and described in the same way across multiple facts and/or data marts, ensuring consistent reporting across the enterprise. A conformed dimension is a set of data attributes that have been physically referenced in multiple database tables using the same key value to refer to the same structure, attributes, domain values, definitions and concepts. A conformed dimension cuts across many facts.

Dimensions are conformed when they are either exactly the same (including keys) or one is a perfect subset of the other. Most important, the row headers produced in two different answer sets from the same conformed dimension(s) must be able to match perfectly. Conformed dimensions are either identical or strict mathematical subsets of the most granular, detailed dimension. Dimension tables are not conformed if the attributes are labeled differently or contain different values. Conformed dimensions come in several different flavors. At the most basic level, conformed dimensions mean exactly the same thing with every possible fact table to which they are joined. The date dimension table connected to the sales facts is identical to the date dimension connected to the inventory facts. For each star schema it is possible to construct fact constellation schema (for example by splitting the original star schema into more star schemes each of them describes facts on another level of dimension hierarchies. The fact constellation architecture contains multiple fact tables that share many dimension tables.

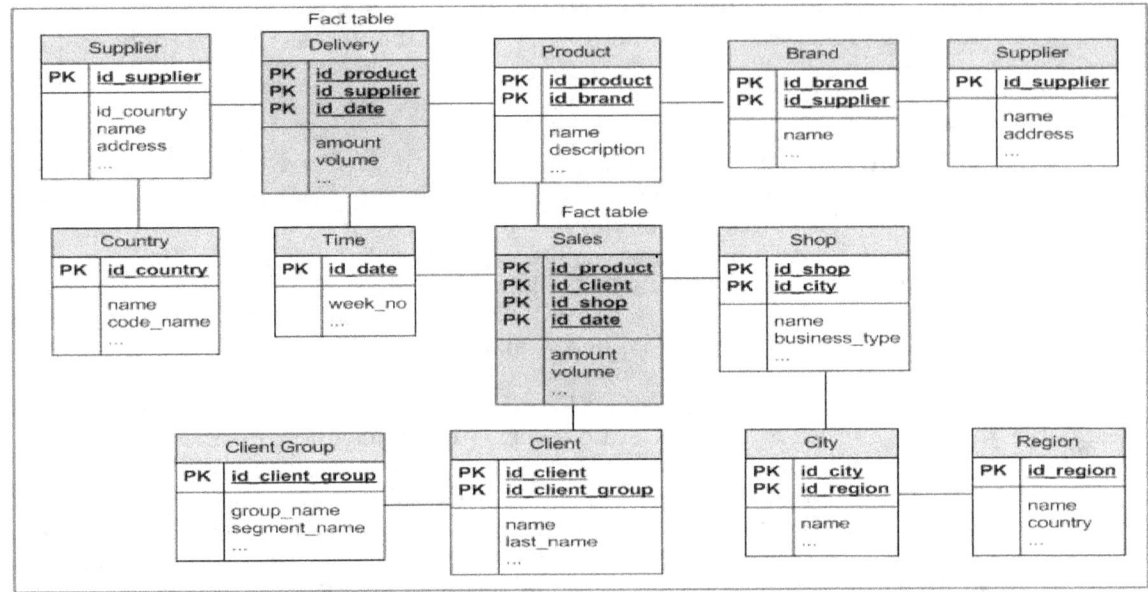

Figure 4.8.1 – Confirmed Dimension

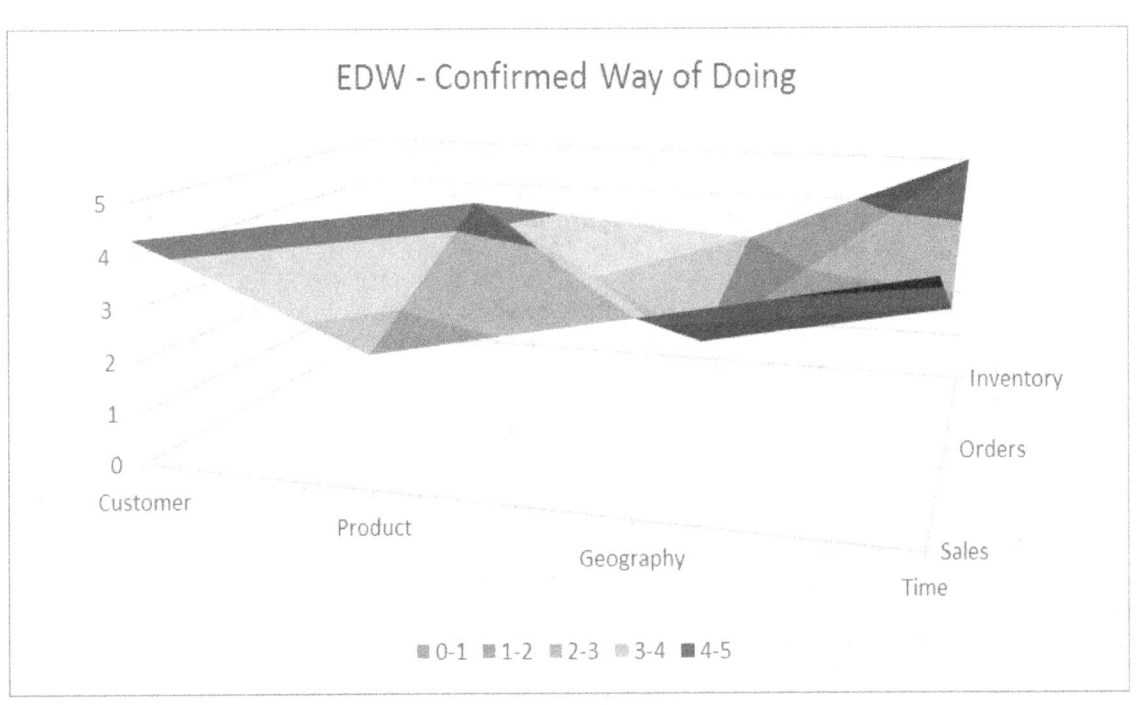

Figure 4.8.2 – Confirmed way of doing business

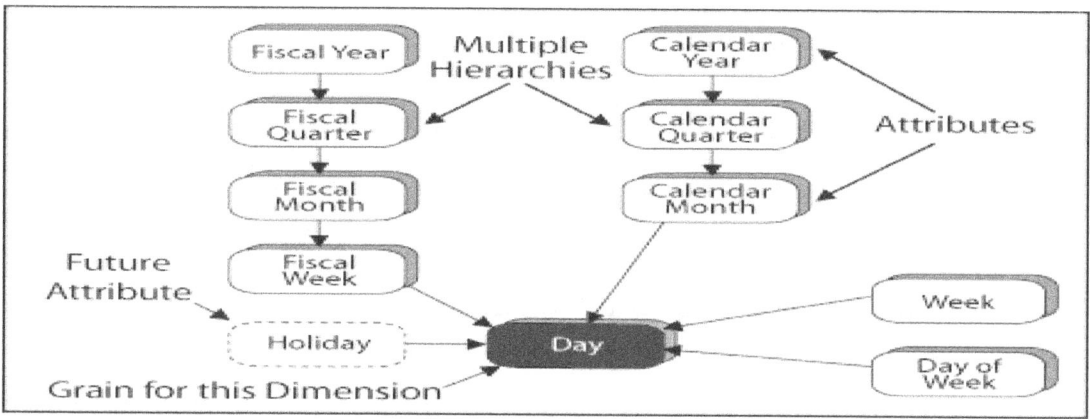

Figure 4.8.3 - Confirmed Dimension

CHAPTER 5
ARCHITECTED DATA MARTS

The evolution is new trend of building architected data marts is to facilitate various frameworks of information processing, management and accessibility on a given data warehousing scope. The architected data marts are thus evolved to follow to follow an architecture for developing the data marts or enterprise data warehouse. There are certain functionalities needs to be built on top of each layer to accommodate more functionality and model driven approach. Any architecture that's needs to be implemented all the once, I a big bang is doomed to failure in today's world. There simply too much risk and too long period to wait until there is payback. A generic migration plan and methodology for development of architected data marts is defined for example to build a corporate data model. All together there is strategy being built in order to achieve the final data warehouse model.

The architected data marts thus have the following characteristics:

- Most timely
- Most accurate
- Most complete
- Nearest to the external source
- Most structurally compatible

The development approach for the data warehouse environment is said to be an iterative or a spiral development approach. The data driven methodology has 3 phases, operational phase, a data warehouse construction phase, and a data warehouse interactive usage phase.

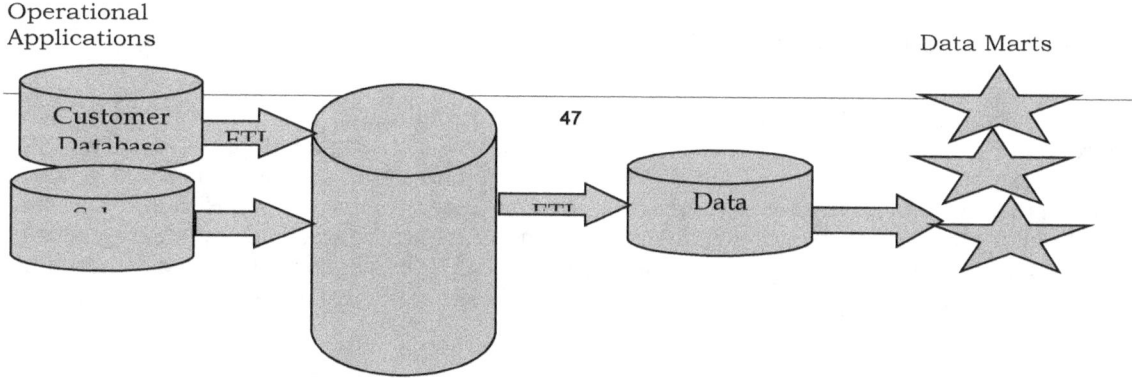

47

Figure 5.1 - Data Warehousing Environment and Top-Down Framework

5.1 Architected Data Marts Supporting the E-Business

A final environment that is supported by the data warehouse is the web-based e-business environment. The data moved from the data warehouse back and forth to web environment. The sorts of things that are done to the data in the Web environment before becoming useful in the data warehouse are the following.

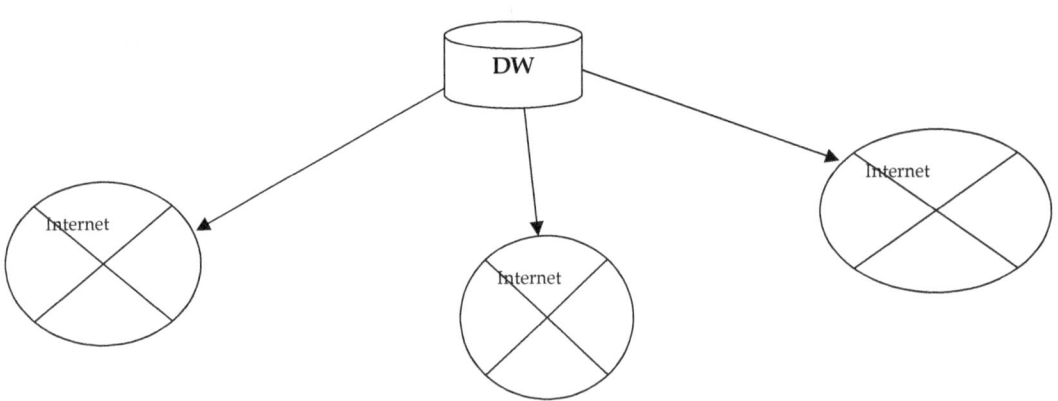

Figure 5.1.1 – The data Warehouse can service more than one e-business

5.2 ERP and Data Warehouse

The basic architecture of the data warehouse, does not change in the face of ERP. The ERP environment contains the application processing where the transactions occur. The ERP integration interface that integrates the data warehouse for non-ERP data. The data warehouse can be built in the ERP environment. As shown in the figure. Here are the data pulled out of the ERP environment and into an operating environment completely unrelated to ERP. The data warehouse can be Oracle.DB2, NT SQL server, or other data warehouse. Once the data leaves the ERP environment, the ERP environment simply becomes another source of data. There are several advantages to pulling data out of the ERP environment and into a different operating environment.

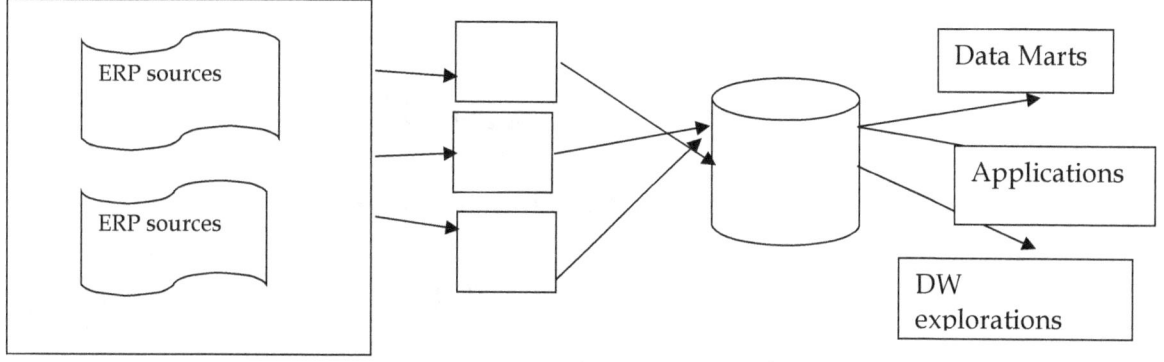

Figure 5.2.1 – The ERP environment feeds the external data warehouse

5.3 The EIS and the Data Warehouse

The EIS is one of the most potent forms of computing. The EIS processing is designed to help the executive make decisions. The EIS becomes the executive's window into the corporation. EIS processing looks across broad vistas and picks out the aspects that are relevant to the running the business. Some of the typical uses of EIS are these:

- Trend analysis
- Key Ration Indicator measurement and tracking
- Drill Down analysis
- Problem Monitoring

- Competitive
- KPI Monitoring

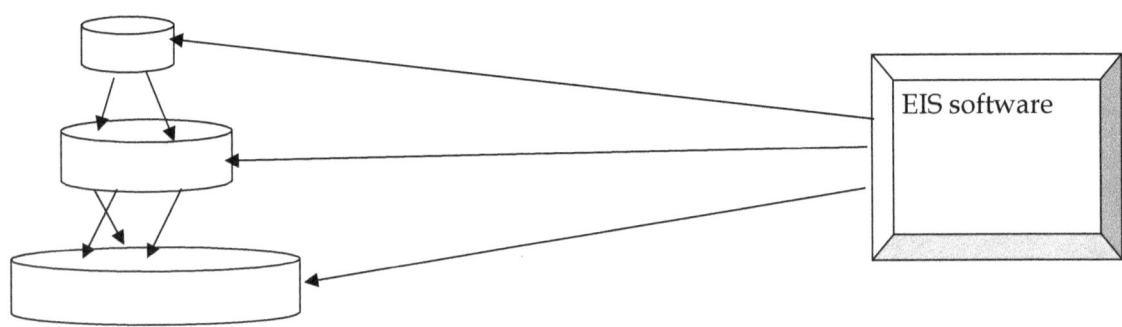

Figure 5.3.1 - EIS and the Data Warehouse

The Architected Data Mart layer consists of the Business Transformation layer, the Reporting layer and the Virtualization layer.

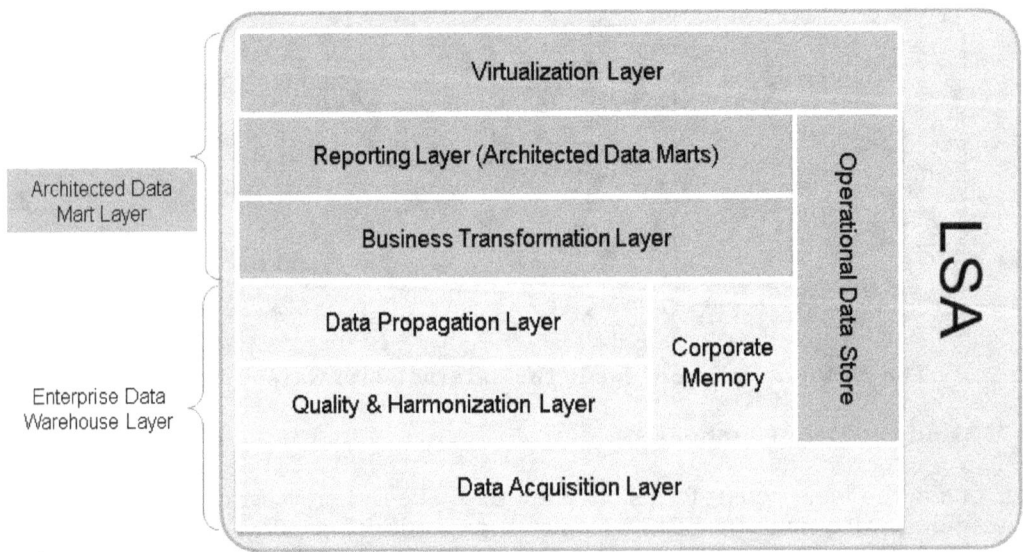

Figure 5.3.2 – Architected Transformation SAP layer

5.3.1 Business Transformation Layer

In the Business Transformation layer, the data is transformed based on business logic. The data in the previous layer (Data Propagation layer) should not be transformed based on business logic, to ensure that the data can be used again. It may be the case that DataStore objects in this layer are needed to compile data from several DataStore objects in the Data Propagation layer.

5.3.2 Reporting Layer

The Reporting layer contains the objects that are used to perform queries for analysis. This layer is modeled mainly using InfoCubes. These cubes save data in BWA. To improve the database performance, you can semantically partition the InfoCubes. Special InfoCubes enable you to create planning scenarios here. Using this InfoCubes as a basis, you can create data views (in the form of aggregation levels) and methods to change data (for example, planning functions and planning sequences). VirtualProviders allow you to access source data directly. Different composite objects (HybridProviders, InfoSets) provide benefits for analysis. Depending on the scenario, you can use these composite InfoProviders.

5.3.3 Virtualization Layer

Queries should always be defined on a Multi Provider for reasons of flexibility. These queries form the Virtualization layer.

5.3.4 Operational Data Store

The Operational Data Store supports operative data analysis. Data can be updated to an operational data store, on a continual basis or in short intervals, and then read for operative analysis. You can also forward the data from the Operational Data Store layer to the data warehouse layer at set times. This means that the data is stored in different levels of granularity. For example, whereas the Operational Data Store layer contains all the changes to the data, only the day-end status is stored in the data warehouse layer.

CHAPTER 6
FEDARATED DATA WAREHOUSING AND DISTRIBUTED DATA WAREHOUSING

Big organization has various regions that provide businesses to customers of that region. Different regional data warehouses were built for each region to meet the specific business needs. A global data warehouse also was built to provide analytical capabilities to executive at global level. The difference between the regional and global data warehouse system is the nature of data resided in each system level. In the regional federated data warehouse architecture picture below, there are two data flows between regional and global data warehouses.

In contrast to capturing and loading data continuously, data federation is another way building data warehouses of de-centralized and federated way. This will reduce the data level accessibility issues of various geographies, times and currencies. Federated data warehouse is a practical approach to achieve the "single version of the truth "across the organization. Federated data warehouse is used to integrate key business measures and dimensions. The foundations of the federated data warehouse are common business model and common staging area. For example a product in sale can be defined as a material in Manufacturing and Equipment in service management. In order integrate these heterogeneous systems that aim to provide the analytical capabilities across the different functions and department, the federated data warehouses are invented.

6.1 Architecture of federated data warehouse

Regional federation possible in federated data warehouse

- Upward federation – only fact data are moved from regional data warehouse to global data warehouse. The aggregation of data can take place at global data warehouse after data integrated or during data movement.
- Downward federation – in downward federation, the reference flows from global to regional level. This ensures the consistency and integrity of data across organization. Transactional data from corporate operational systems such as ERP, CRM... are sourced at global level and then extracted, transformed and loaded into respective regional data warehouse

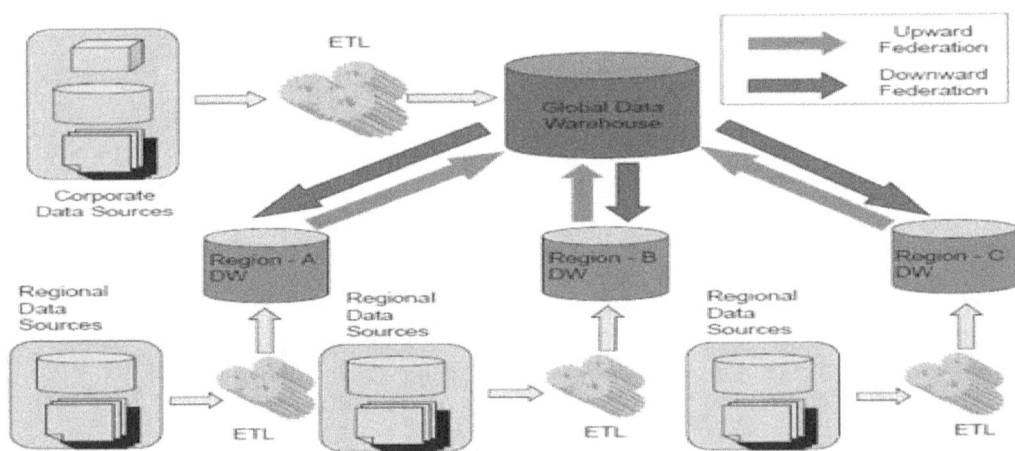

Figure 6.1 – Regional Federation- Federated Data Warehouse

6.2 Functional Federation possible in federated data warehouse

Functional federated data warehouse is used when the organizations has different data warehouses system was built for specific applications such as ERP, CRM or subject specific. The components of functional federated data warehouse architecture includes data marts, custom built data warehouses, ETL tools, cross function reporting systems, real time data store and reporting as picture below:

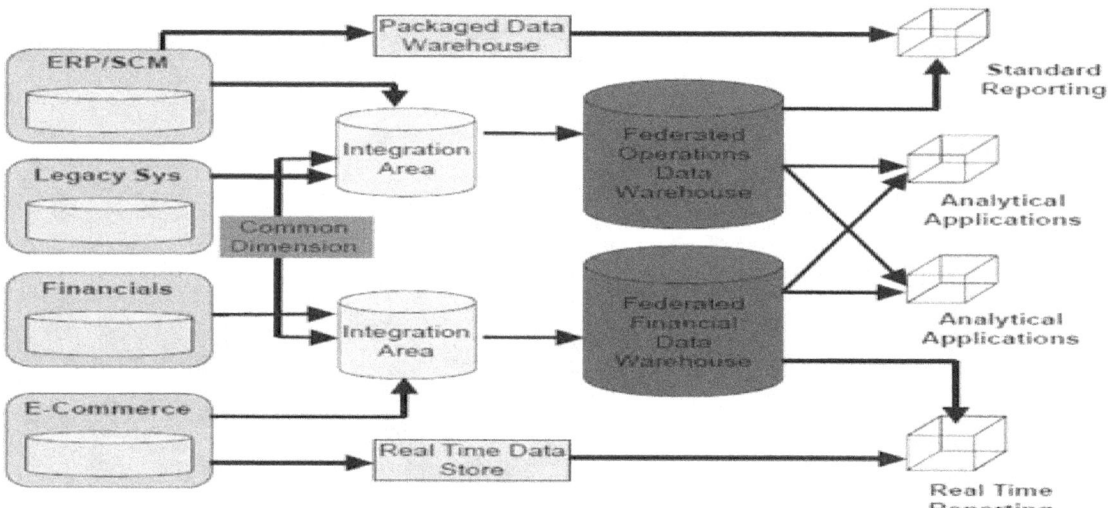

Figure 6.2 – Functional Federation – Federated Data Warehouse

6.3 Benefits of Data Warehouse

- Ease of implementation – Federated data warehouse integrated all legacy data warehouses, business intelligence systems into a newer system that provides analytical capabilities across function. Federated data warehouse data not try to rebuild a new system which potentially causes major point of conflict
- Query Optimization and set operations
- Shorter implementation time – By integrating all legacy BI systems, federated data warehouse approach has a shorter implementation time in compare with lengthy processing of building enterprise data warehouse.

- Cross functional analytics requirements – Cross function analytics requirements accomplished using common business module across different BI systems of each department. Federated data warehouse is a dynamic cooperation of various business intelligence systems to make them talk to each other and cube support
- Business is distributed geographically or over multiple

Federated data warehouse offers a practical solution for building data warehouse. The iterative manner of federated data warehouse approach helps to reduce the implementation time and cost and therefore provide excellent proposition to the business.

CHAPTER 7
REAL TIME DATA WAREHOUSING

For the discussion under review, retrofitting RTDW to a traditional, history oriented DW is complex and expensive as is upgrading or replacing a database management system to get a platform more conducive to RTDW, namely the intelligence, analytics and broader view of business entries that operational processes and operational applications gain from tight integration with a data warehouse in a near real time operational environment. Real-time and similar functions are key enablers for RTDW. As it process the data, in-line and make it available as when required or change based, analytics power of the data warehouse and related business intelligence infrastructure. There are many frame works and models to support and enhance the capability of real time operational data, such as IBM specialized CDC and also with many ETL tools.

Real-time business intelligence (RTBI) is the process of delivering business intelligence or information about business operations as they occur. Real time means near to zero latency and access to information whenever it is required. The speed of today's processing systems has moved classical data warehousing into the realm of real-time. The result is real-time business intelligence. Business transactions as they occur are fed to a real-time BI system that maintains the current state of the enterprise. The RTBI system not only supports the classic strategic functions of data warehousing for deriving information and knowledge from past enterprise activity, but it also provides real-time tactical support to drive enterprise actions that react immediately to events as they occur. As such, it replaces both the classic data warehouse and the enterprise application integration (EAI) functions. Such event-driven processing is a basic tenet of real-time business intelligence. Many vendors derived various technologies for driving the data integration for processing change data captures, or even the enterprise process integration like Bigdata message ques, have massive volumes like 60k hit per second or even more of converting and capturing the machine learning paradigms.

In this context, "real-time" means a range from milliseconds to a few seconds after the business event has occurred. While traditional BI presents historical data for manual analysis, RTBI compares current business events with historical patterns to detect problems or opportunities automatically. This automated analysis capability enables corrective actions to be initiated and/or business rules to be adjusted to optimize business processes. RTBI is an approach in which up-to-a-minute data is analyzed, either directly from Operational sources or feeding business transactions into a real time data warehouse and Business Intelligence system. RTBI analyzes real time data. Real-time business intelligence makes sense for some applications but not for others – a fact that organizations need to take into account as they consider investments in real-time BI tools. Key to deciding whether a real-time BI strategy would pay dividends is understanding the needs of the business and determining whether end users require immediate access to data for analytical purposes, or if something less than real time is fast enough.

7.1 Evolution of RTBI

In today's competitive environment with high consumer expectation, decisions that are based on the most current data available to improve customer relationships, increase revenue, maximize operational efficiencies, and yes – even save lives. This technology is real-time business intelligence.   Real-time business intelligence systems provide the information necessary to strategically improve an enterprise's processes as well as to take tactical advantage of events as they occur.

7.1.1 Latency

All real-time business intelligence systems have some latency, but the goal is to minimize the time from the business event happening to a corrective action or notification being initiated. Analyst Richard Hackathorn describes three types of latency:

- Data latency; the time taken to collect and store the data
- Analysis latency; the time taken to analyze the data and turn it into actionable information
- Action latency; the time taken to react to the information and take action

Real-time business intelligence technologies are designed to reduce all three latencies to as close to zero as possible, whereas traditional business intelligence only seeks to reduce data latency and does not address analysis latency or action latency since both are governed by manual processes. Some commentators have introduced the concept of right time business intelligence which proposes that information should be delivered just before it is required, and not necessarily in real-time.

7.1.2 Event-based

Real-time Business Intelligence systems are event driven, and may use Complex Event Processing, Event Stream Processing and Mashup (web application hybrid) techniques to enable events to be analyzed without being first transformed and stored in a database. These in- memory techniques have the advantage that high rates of events can be monitored, and since data does not have to be written into databases data latency can be reduced to milliseconds.

7.1.3 Data Warehouse

An alternative approach to event driven architectures is to increase the refresh cycle of an existing data warehouse to update the data more frequently. These real-time data warehouse systems can achieve near real-time update of data, where the data latency typically is in the range from minutes to hours. The analysis of the data is still usually manual, so the total latency is significantly different from event driven architectural approaches.

7.1.4 Server-less technology

The latest alternative innovation to "real-time" event driven and/or "real-time" data warehouse architectures is MSSO Technology (Multiple Source Simple Output) which removes the need for the data warehouse and intermediary servers altogether since it is able to access live data directly from the source (even from multiple, disparate sources). Because live data is accessed directly by server-less means, it provides the potential for zero-latency, real-time data in the truest sense.

7.1.5 Process-aware

This is sometimes considered a subset of Operational intelligence and is also identified with Business Activity Monitoring. It allows entire processes (transactions, steps) to be monitored, metrics (latency, completion/failed ratios, etc.) to be viewed, compared with warehoused historic data, and trended in real-time. Advanced implementations allow threshold detection, alerting and providing feedback to the process execution systems themselves, thereby 'closing the loop'.

7.1.6 Technologies support Real-Time Analytics

Technologies that can be supported to enable real-time business intelligence are data visualization, data federation, enterprise information integration, enterprise application integration and service oriented architecture. Complex event processing tools can be used to analyze data streams in real time and either trigger automated actions or alert workers to patterns and trends.

Data warehouse appliance: Data warehouse appliance is a combination of hardware and software product which was designed exclusively for analytical processing. In data warehouse implementation, tasks that involve tuning, adding or editing structure around the data, data migration from other databases, reconciliation of data are done by DBA. Another task for DBA was to make the database to perform well for large sets of users. Whereas with data warehouse appliances, it is the vendor responsibility of the physical design and tuning the software as per hardware requirements. Data warehouse appliance package comes with its own operating system, storage, DBMS, software, and required hardware. If required data warehouse appliances can be easily integrated with other tools.

Mobile technology: There are very limited vendors for providing Mobile business intelligence; MBI is integrated with existing BI architecture. MBI is a package that uses existing BI applications so people can use on their mobile phone and make informed decision in real time.

7.1.7 Application Areas

* Arithmetic Algorithm
* Fraud detection
* Systems monitoring
* Application performance monitoring
* Customer Relationship Management
* Demand sensing
* Dynamic pricing and yield management
* Data validation
* Operational intelligence and risk management
* Payments & cash monitoring
* Data security monitoring
* Supply chain optimization
* RFID/sensor network data analysis
* Work Streaming
* Call centre optimization
* Enterprise Mashups and Mashup Dashboards
* Transportation industry

Case Study: Transportation industry can be benefited by using real-time analytics. For an example railroad network. Depending on the results provided by the real-time analytics, dispatcher can make a decision on what kind of train he can dispatch on the track depending on the train traffic and commodities shipped.

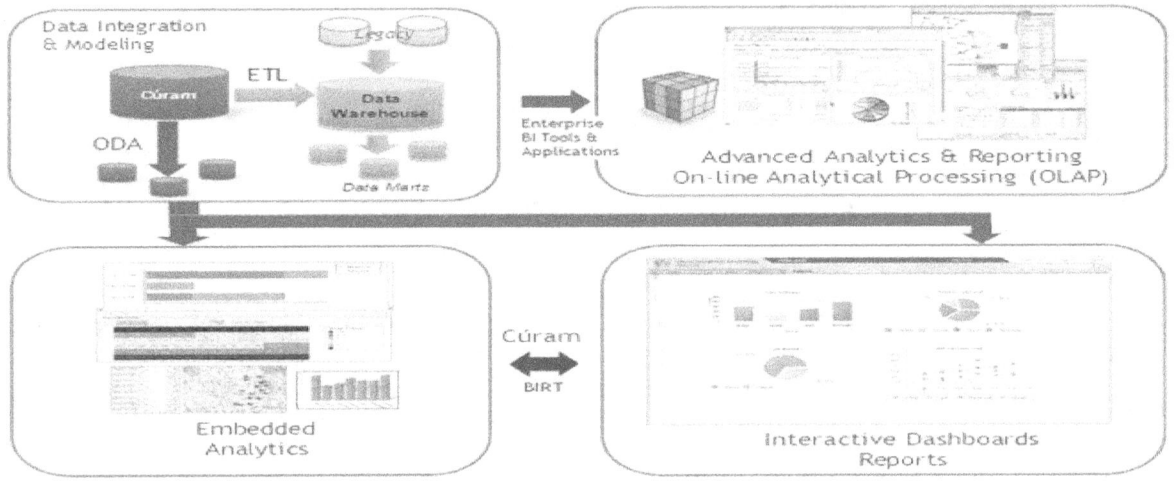

Figure 7.1 - Real Time Data Warehousing Implementation

7.1.8 Implementation of Big data message qs and click streams for RTBI

IBM MQ one of the message processing techniques where **it** can transport any type of data as messages, enabling businesses to build flexible, reusable architectures such as service-oriented architecture (SOA) and self BI environments. It works with broad range of computing platforms, applications, web services and communications **protocols for security-rich message delivery.** IBM MQ acts as a middle layer for visibility and control of the flow of messages and data inside and outside your organization. IBM MQ is messaging and queuing middleware, with several modes of operation: point-to-point, publish/subscribe; file transfer. These applications can publish messages to many subscribers.

7.1.9 Oracle Real Application clusters (RAC) and flash back recovery

Shared-nothing architectures share neither the data on disk nor the data in memory between nodes in the cluster.

CHAPTER 8
MASTER DATA MANAGEMENT

MDM ensures better integration by identifying data properly, so that the best sources are found better integration by identifying data properly. So that the best sources are found for a specific use and data exchanges are made. An MDM framework would bring-up lots of discipline, in terms of data management, customer 360 view, organizational handling of information, in-house availability of the data for the downstream applications and also reduce the DQ functions.

- Match merge process
- Record linkage
- Data Reconciliation

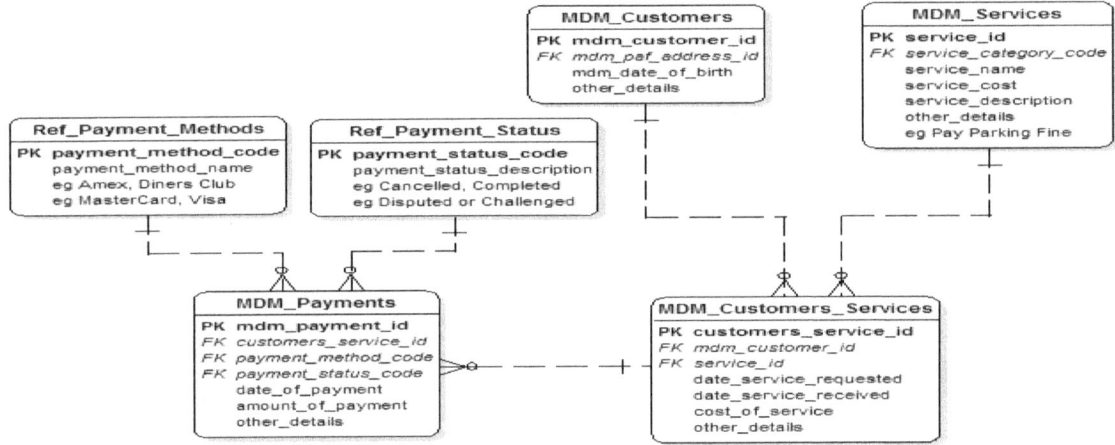

Figure 8.1.1 – MDM customer data

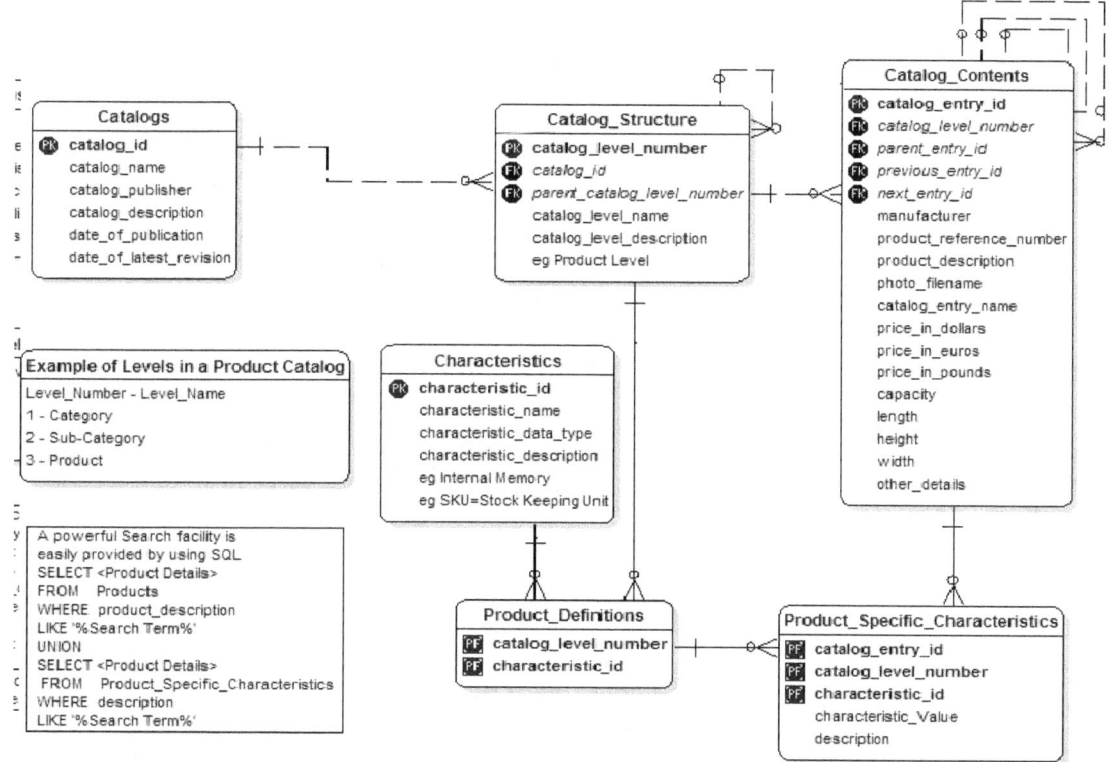

Figure 8.1.2 – MDM Product Data

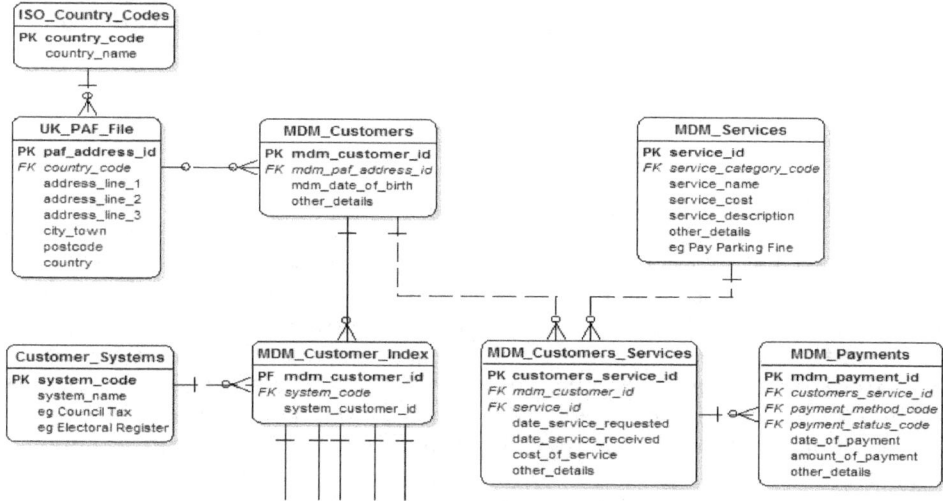

Figure 8.1.3 – Reference data MDM

CHAPTER 9
IN MEMERY PROCESSING AND 64- BIT COMPUTING

In short, in-memory processing enables users to have immediate access to the right Information which results in more informed decisions. With traditional Business Intelligence (BI) technology, data is loaded onto disks in the form of tables and multi-dimensional cubes against which queries are run. While In-memory processing has a great potential for end users it is not the answer to everyone. Important question organizations need to ask is if slower query response times are preventing users from making important decisions. If company is a slow moving business where things don't change often then in-memory solution is not effective. Organizations where there is a significant growth in data volume and increase in demand for reporting functionalities that facilitate new opportunities would be a right scenario to deploy in-memory BI. Security needs to be the first and foremost concern when deploying In-memory tools as they expose huge amounts of data to end users. Care should be taken as to who has access to the data, how and where data is stored. End users download huge amounts of data onto their desktops and there is danger of data getting compromised. It could get lost or stolen. Measures should be taken to provide access to the data only to authorized users. The arrival of column centric databases which stored similar information together allowed storing data more efficiently and with greater compression. This in turn allowed to store huge amounts of data in the same physical space, reducing the amount of memory needed to perform a query and increasing processing speed. With in-memory database, all information is initially loaded into memory. This eliminates the need for optimized databases, indexes, aggregates and designing of cubes star schemas.

Most in-memory tools use compression algorithms that reduce the size of in-memory beyond what would be necessary for hard disks. Users query the data loaded into the system's memory thereby avoiding slower database access and performance bottlenecks. This differs from caching, a very widely used method to speed up query performance, in that caches are subsets of very specific pre-defined organized data. With in-memory tools, data available for analysis can be as large as data mart or small data warehouse which is entirely in memory. This can be accessed within seconds by multiple concurrent users at a detailed level and offers the potential for excellent analytics. Theoretically the improvement in data access is 10,000 to 1,000,000 times faster than from disk. It also minimizes the need for performance tuning by IT staff and provides faster service for end users.

9.1 Cheaper and higher performing hardware

According to Moore's law the computing power doubles every two to three years while decreasing in costs. CPU processing, memory and disk storage are all subject to some variation of this law. Also hardware innovations like multi-core architecture, NAND flash memory, parallel servers, increased memory processing capability, etc. and software innovations like column centric databases, compression techniques and handling aggregate tables, etc. have all contributed to the demand of In-memory products. As the data used by organizations grew traditional data warehouses just couldn't deliver a timely, accurate and real time data. The extract, transform, load (ETL) process that periodically updates data warehouses with operational data can take anywhere from a few hours to weeks to complete. So at any given point of time data is at least a day old. In-memory processing makes easy to have instant access to terabytes of data for real time reporting. In-memory processing comes at a lower cost and can be easily deployed and maintained when compared to traditional BI tools. According to Gartner survey deploying traditional BI tools can take as long as 17 months. Many data warehouse vendors are choosing In-memory technology over traditional BI to speed up implementation times.

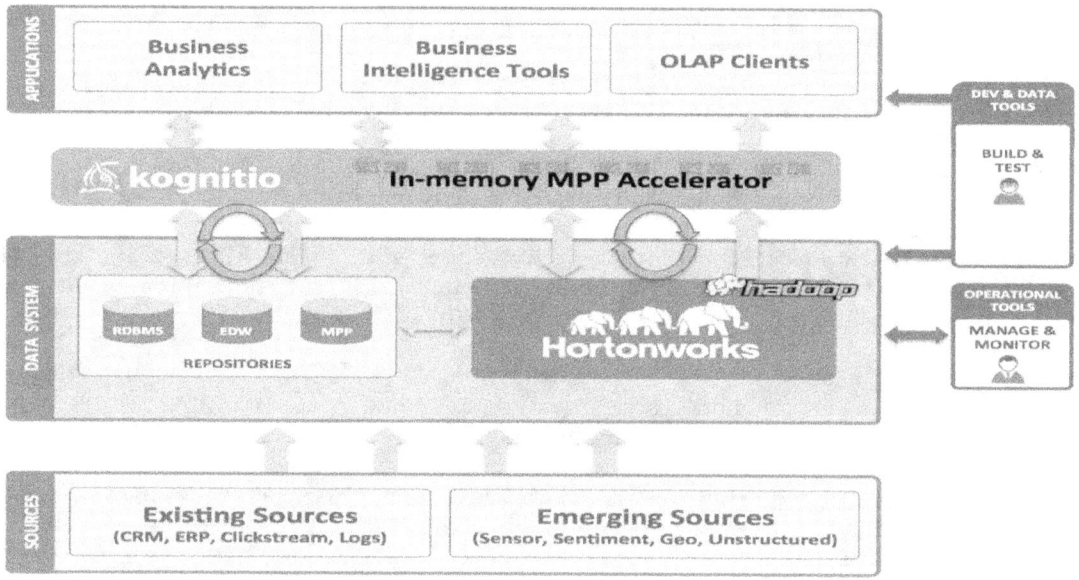

Figure 9.1.1 – In-memory computing

9.2 Advantages of In-memory BI

Several in-memory vendors provide ability to connect to existing data sources and access to visually rich interactive dashboards. This allows business analysts and end users to create custom reports and queries without much training or expertise. Easy navigation and ability to modify queries on the fly is an appealing factor to many users. Since these dashboards can be populated with fresh data, it allows users to have access to real time data and create reports within minutes, which is a critical factor in any business intelligence application.

With In-memory processing the source database is queried only once instead of accessing the database every time a query is run thereby eliminating repetitive processing and reducing the burden on database servers. By scheduling to populate In-memory database overnight the database servers can be used for operational purposes during peak hours. In-memory processing can be a blessing in disguise for operational workers such as call centre representatives or warehouse managers who need instant and accurate data to make fast decisions.

9.3 64-BIT COMPUTING

Though the idea of In-memory technology is not new, it is only recently emerging thanks to the widely popular and affordable 64-bit processors and declining memory chips prices. 64 bit operating systems allows access to far more RAM (up to 100GB or more) than the 2 or 4 GB accessible on 32-bit systems. By providing Terabytes (1 TB = 1,024 GB) of space available for storage and analysis, 64-bit operating systems make in-memory processing scalable. The use of flash memory enables systems to scale to many Terabytes more economically. In computer architecture, 64-bit computing is the use of processors that have data path widths, integer size, and memory address widths of 64 bits (eight octets). Also, 64-bit CPU and ALU architectures are those that are based on registers, address, or data buses of that size. From the software perspective, 64-bit computing means the use of code with 64-bit virtual memory addresses. The term 64-bit describes a generation of computers in which 64-bit processors are the norm. 64 bits is a word size that defines certain classes of computer architecture, buses, memory and CPUs, and by extension the software that runs on them. 64-bit CPUs have been used in supercomputers since the 1970s (Cray-1, 1975) and in RISC based workstations and servers since the early 1990s, notably the MIPS R4000, R8000, and R10000, the DEC Alpha, the Sun Ultra SPARC, and the IBM RS64 and POWER3 and later POWER microprocessors. In 2001 NEC released a 64 bit RISC CPU for mobile devices, notably the low cost Casio BE-300/ In 2003 64-bit CPUs were introduced to the (previously 32-bit) mainstream personal computer arena in the form of the x86-64 and 64-bit PowerPC processor architectures and in 2012[3] even into the ARM architecture targeting smartphones and tablet computers, first sold on September 20, 2013 in the iPhone 5S powered by the ARMv8-A Apple A7 SoC.

A 64-bit register can store 264 (over 18 quintillion or 1.8×1019) different values. Hence, a processor with 64-bit memory addresses can directly access 264 bytes of byte addressable memory. Without further qualification, a 64-bit computer architecture generally has integer and addressing registers that are 64 bits wide, allowing direct support for 64-bit data types and addresses. However, a CPU might have external data buses or address buses with different sizes from the registers, even larger (the 32 bit Pentium had a 64-bit data bus, for instance. The term may also refer to the size of low-level data types, such as 64-bit floating-point numbers.

Processor registers are typically divided into several groups: integer, floating-point, SIMD, control, and often special registers for address arithmetic which may have various uses and names such as address, index or base registers. However, in modern designs, these functions are often performed by more general purpose integer registers. In most processors, only integer or address-registers can be used to address data in memory; the other types of registers cannot. The size of these registers therefore normally limits the amount of directly addressable memory, even if there are registers, such as floating-point registers, that are wider. Most high performance 32-bit and 64-bit processors (some notable exceptions are older or embedded ARM and 32-bit MIPS CPUs) have integrated floating point hardware, which is often, but not always, based on 64-bit units of data. For example, although the x86/x87 architecture has instructions capable of loading and storing 64-bit (and 32-bit) floating-point values in memory, the internal floating point data and register format is 80 bits wide, while the general-purpose registers are 32 bits wide. In contrast, the 64-bitAlpha family uses a 64-bit floating-point data and register format.

Table 9.1 64-bit data models

LLP64/ IL32P64	16	32	32	64	64	Microsoft Windows (x86-64 and IA-64)
LP64/ I32LP64	16	32	64	64	64	Most Unix and Unix-like systems, e.g. Solaris, Linux, BSD, and OS X; z/OS
ILP64	16	64	64	64	64	HAL Computer Systems port of Solaris to SPARC64
SILP64	64	64	64	64	64	"Classic" UNICOS (as opposed to UNICOS/mp, etc.)

In 32-bit programs, pointers and data types such as integers generally have the same length; this is not necessarily true on 64-bit machines. Mixing data types in programming languages such as C and its descendants such as C++ and Objective-C may thus function on 32-bit implementations but not on 64-bit implementations. In many programming environments for C and C-derived languages on 64-bit machines, "int" variables are still 32 bits wide, but long integers and pointers are 64 bits wide. These are described as having an LP64 data model. Another alternative is the ILP64 data model in which all three data types are 64 bits wide, and even SILP64 where "short" integers are also 64 bits wide. However, in most cases the modifications required are relatively minor and straightforward, and many well-written programs can simply be recompiled for the new environment without changes. Another alternative is the LLP64 model, which maintains compatibility with 32-bit code by leaving both int and long as 32-bit. "LL" refers to the "long long integer" type, which is at least 64 bits on all platforms, including 32-bit environments.

Many 64-bit platforms today use an LP64 model (including Solaris, AIX, HP-UX, Linux, OS X, BSD, and IBM z/OS). Microsoft Windows uses an LLP64 model. The disadvantage of the LP64 model is that storing a long into an int may overflow. On the other hand, converting a pointer to a long will "work" in LP64. In the LLP64 model, the reverse is true. These are not problems which affect fully standard-compliant code, but code is often written with implicit assumptions about the widths of data types. C code should prefer (u) intptr_t instead of long when casting pointers into integer objects. Note that a programming model is a choice made on a per-compiler basis, and several can coexist on the same OS. However, the programming model chosen as the primary model for the OS API typically dominates. Another consideration is the data model used for drivers. Drivers make up the majority of the operating system code in most modern operating system. Although many may not be loaded when the operating system is running). Many drivers use pointers heavily to manipulate data, and in some cases have to load pointers of a certain size into the hardware they support for DMA. As an example, a driver for a 32-bit PCI device asking the device to DMA data into upper areas of a 64-bit machine's memory could not satisfy requests from the operating system to load data from the device to memory above the 4 gibibyte barrier, because the pointers for those addresses would not fit into the DMA registers of the device. This problem is solved by having the OS take the memory restrictions of the device into account when generating requests to drivers for DMA, or by using an IOMMU.

Table 9.2 Sample 64 bit Architecture

Table 9.2.1 - IBM Power 6 Servers

Name	Number of sockets	Number of cores	CPU clock frequency
520 Express	2	4	4.2 GHz or 4.7 GHz
550 Express	4	8	4.2 GHz or 5.0 GHz
560	8	16	3.6 GHz

Name	Number of sockets	Number of cores	CPU clock frequency
Express			
570	8	16	4.4 GHz or 5.0 GHz
570	16	32	4.2 GHz
575	16	32	4.7 GHz
595	32	64	4.2 GHz or 5.0 GHz

Table- 9.2.2 -IBM POWERS 6 blade servers

Name	Number of cores	CPU clock frequency	Blade slots required
BladeCenter JS12	2	3.8 GHz	1
BladeCenter JS22	4	4.0 GHz	1
BladeCenter JS23	4	4.2 GHz	1
BladeCenter JS43	8	4.2 GHz	2

CHAPTER 10
OPEN SOURCE DATA WAREHOUSING AND BUSINESS INTELLIGENCE

Open Source Data Warehousing and Business Intelligence is an all-in-one reference for developing open source based data warehousing (DW) and business intelligence (BI) solutions that are business-centric, cross-customer viable, cross-functional, cross-technology based, and enterprise-wide. Considering the entire lifecycle of an open source DW & BI implementation, highlighting the key differences between open source and vendor DW and BI technologies, this book identifies end-to-end solutions that are scalable, high performance, and stable. It illustrates the practical aspects of implementing and using open source DW and BI technologies to supply with valuable on-the-project experience that can help you improve implementation and productivity. Emphasizing analysis, design, and programming, the text explains best-fit solutions as well as how to maximize ROI. Coverage includes data warehouse design, real-time processing, data integration, presentation services, and real-time reporting. With a focus on real-world applications. Open source not only delivered low or no-cost, liberally licensed software that opened up capabilities to even the smallest companies, but also opened up code and functionality to a community process that ensured solutions remained true to prevailing standards. In the process, solutions became far more straightforward and flexible to implement. The open source data warehouse employs the same licensing model, the same community development process and same degree of openness as other types of open source software. Simply put, most leading open source products are offered either as a free download, or for a nominal fee, as a fully supported system. In either case, there is no limit to the number of licenses and implementations a company may make with the software; and users have large, committed communities they can turn to for additional support or upgrades.

Highlighting the key differences between open source and vendor DW and BI technologies:
- The open source data warehouses reduces the dependency on the Vendor evaluation and comes upfront, provides less cost for the support and maintenance.

- Details the practical aspects of open source data warehouse and business Intelligence Architecture and technologies
- Supplies on-the-project experience that helps to improve implementation and Productivity
- Presents best practices in data management and application management

- Emphasizes analysis, design, and programming

- Explains best-fit solutions as well as how to maximize ROI

- The Open Source BI architecture delivers a more controlled and better approach for enterprise wide solution

- Increased adoption, reliability and more choices drive the transition due to processing of small amounts of data volumes

10.1 Open Source DW & BI: Successful Players and Products

Oracle: MySQL Vendor
PostgreSQL Vendor
Infobright
Pentaho: Mondrian Vendor
Jedox: Palo Vendor
Dynamo BI and Eigenbase: LucidDB Vendor
GreenPlum Vendor
HadoopDB
Talend Data Integration   Actuate BIRT BI Platform
JasperSoft Enterprise
Pentaho Enterprise BI Suite
KNIME (Konstanz Information Miner)
Pervasive DataRush
Pervasive DataCloud2

Data Warehousing and Business Intelligence: An Open Source Solution

Data warehousing solution for small scale retail and banking customers

Open Source Oracle SQL Vendor

It offers up to 1 TB of data along with the ability to transform CSV data and Excel sheets.

Oracle: MySQL Vendor

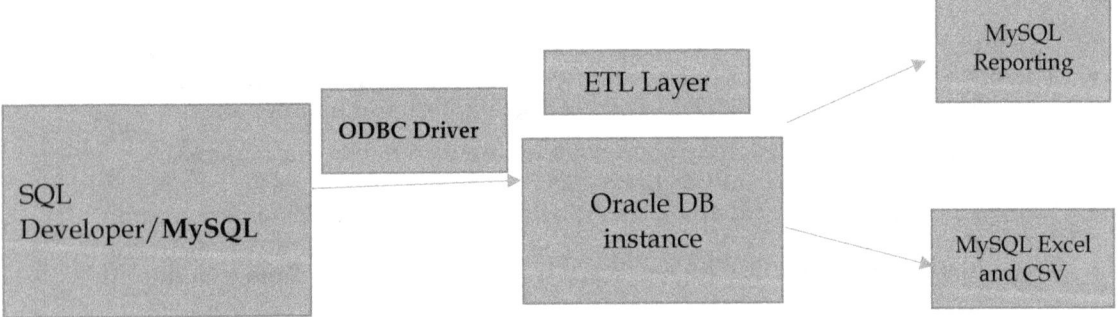

CHAPTER 11
COLUMNAR DATA WAREHOUSING

The past decade has been rapid evolution of the Data Warehouse with row based algorithm and distribution based on SOA and nodes lead to challenges with respect to data compression, adoptability and focus on reducing the need for reverse-key indexing, enhancing the power of building the analytics based on the vertical table column structures .The business landscape and business organizations are increasingly realizing the need for more scalable and flexible information technology architecture.

A column-oriented DBMS is a database management system (DBMS) that stores data tables as sections of columns of data rather than as rows of data. In comparison, most relational DBMSs store data in rows. This column-oriented DBMS has advantages for data warehouses, customer relationship management (CRM) systems, and library card catalogs, and other ad hoc inquiry systems where aggregates are computed over large numbers of similar data items. It is possible to achieve some of the benefits of column-oriented and row-oriented organization with any DBMSs. Denoting one as column-oriented refers to both the ease of expression of a column-oriented structure and the focus on optimizations for column-oriented workloads. This approach is in contrast to row-oriented or row store databases and with correlation databases, which use a value-based storage structure. Column-oriented storage is closely related to database normalization due to the way it restricts the database schema design. However, it was often found to be too restrictive in practice, and thus many column-oriented databases such as Google's BigTable do allow "column groups" to avoid frequently needed joins.

11.1 Benefits Columnar Databases

Comparisons between row-oriented and column-oriented data layouts are typically concerned with the efficiency of hard-disk access for a given workload, as seek time is incredibly long compared to the other delays in computers. Sometimes, reading a megabyte of sequentially stored data takes no more time than one random access. Further, because seek time is improving much more slowly than CPU power (see Moore's Law), this focus will likely continue on systems that rely on hard disks for storage. Following is a set of oversimplified observations which attempt to paint a picture of the trade-offs between column- and row-oriented organizations. Unless, of course, the application can be reasonably assured to fit most/all data into memory, in which case huge optimizations are available from in-memory database systems.

In practice, row-oriented storage layouts are well-suited for **OLTP**-like workloads which are more heavily loaded with interactive transactions. Column-oriented storage layouts are well-suited for **OLAP**-like workloads (e.g., **data warehouses**) which typically involve a smaller number of highly complex queries over all data (possibly **terabytes**).

11.2 Data Compression

Column data is of uniform type; therefore, there are some opportunities for storage size optimizations available in column-oriented data that are not available in row-oriented data.

For example, many popular modern compression schemes, such as **LZW** or **run-length encoding**, make use of the similarity of adjacent data to compress. While the same techniques may be used on row-oriented data, a typical implementation will achieve less effective results.

To improve compression, sorting rows can also help. For example, using bitmap indexes, sorting can improve compression by an order of magnitude. To maximize the compression benefits of the **lexicographical order** with respect to **run-length encoding**, it is best to use low-cardinality columns as the first sort keys. For example, given a table with columns sex, age, name, it would be best to sort first on the value sex (cardinality of two), then age (cardinality of <150), then name. Columnar compression achieves a reduction in disk space at the expense of efficiency of retrieval. Retrieving all data from a single row is more efficient when that data is located in a single location, such as in a row-oriented architecture. Further, the greater adjacent compression achieved, the more difficult random-access may become, as data might need to be uncompressed to be read. Therefore, column-oriented architectures are sometimes enriched by additional mechanisms aimed at minimizing the need for access to compressed data.

11.3 Implementations

While even a traditional row-oriented RDBMS system can achieve some benefits of column-oriented layout, specialization of the storage layer and of the query-execution engine provide further benefits. While nothing precludes providing both row- and column-optimized capabilities in a single DBMS, typically products specialize in one of these directions.

11.4 List of column oriented DBMS

DB
Calpont InfiniDB,Druid,MonetDB,RCFile

IaaS

1010 data, Amazon RedShift, Google Big Query

Proprietary

Endeca,EXASOL,Greenplum Database, IBM DB2,Infobright,KDB,memSQL,ParAccel,SenSage, SAP HANA,Microsoft SQL Server 2012,Sybase IQ,Teradata, Vector- formerly Vectorwise, Vertica (developed from open-source C-Store), Actuate Corporation BIRT Analytics Columnar DB

Relation

A	B	C	D
101	111	121	131
102	112	122	132
103	113	123	133
104	114	124	134
105	115	105	135

RCFile

...
...
HDFS Blocks
Row Group 1
Row Group 2
...
Row Group n
...

Row Group

16 Bytes Sync	Metadata Header			
101	102	103	104	105
111	112	113	114	115
121	122	123	124	125
131	132	133	134	135

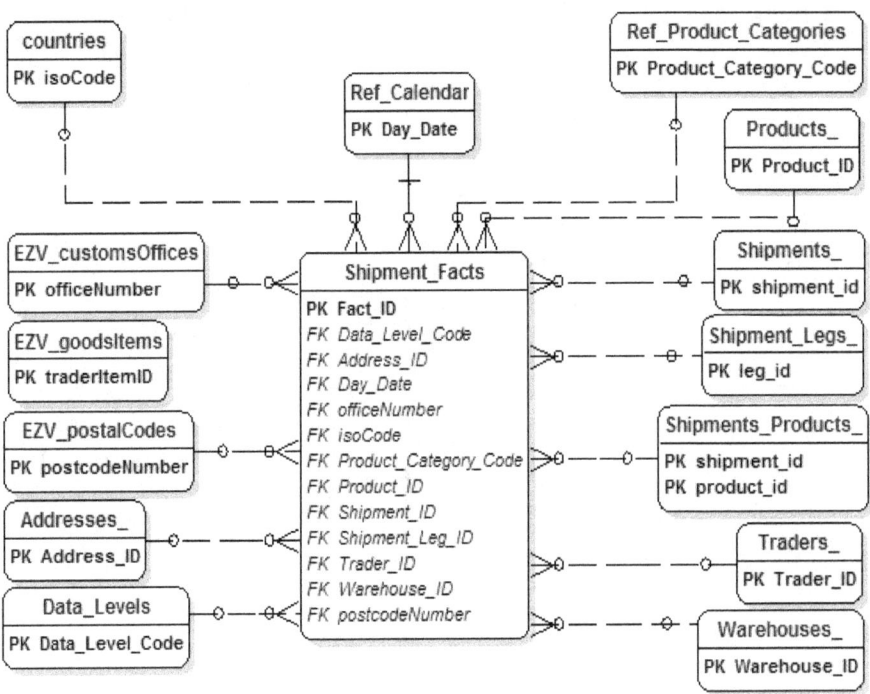

Figure 11.1 - Columnar Structures

CHAPTER 12
DATA WAREHOUSE APPLIANCES AND SOFTWARE APPLIANCES

Many Data Warehouse appliances and software appliances were developed late during 2003 and 2006 to accommodate challenging data growth and also to improve the performance and reliability as ready to use systems as against the traditional I/O or CPU based pricing models. Partitioning and indexing options have become more and more inaccurate and complicated to obtain any new approach. There are many data warehouse appliances are in the market like Oracle Exadata from Oracle, Green Plum from Sun Micro Systems, DATA Allegro partnered with Microsoft and Teradata as usual as conventional MPP to support appliance usage.

The data warehouse appliance (DWA) has several characteristics which differentiate that architecture from similar machines in a data center, such as an enterprise data warehouse (EDW). A DWA has a very tight integration of its internal components which are optimized for "data-centric" operations in contrast to "compute-centric" operations. The latter tend to emphasize number of CPU's, cores and network bandwidth. A DWA is trivial to use and install. In contrast to a "pre-configuration" of components, a DWA has very few configuration switches or options. The elimination of such options significantly reduces configuration error – the number one cause for failure in large systems. A DWA is optimized for analytics on Big data. In contrast, preceding architectures (including parallel ones) focused on "enterprise data warehouse" being a general-purpose repository for data and supporting analytics as an ancillary task. A DWA has high performance for analytics on big data. The price-performance is usually more than 50 to 100X that of earlier architectures such as EDW. Most DW appliances use massively parallel processing (MPP) architectures to provide high query performance and platform scalability. MPP architectures consist of independent processors or servers executing in parallel. Most MPP architectures implement a "shared-nothing architecture" where each server operates self-sufficiently and controls its own memory and disk. DW appliances distribute data onto dedicated disk storage units connected to each server in the appliance. This distribution allows DW appliances to resolve a relational query by scanning data on each server in parallel. The divide-and-conquer approach delivers high performance and scales linearly as new servers are added into the architecture.

12.1 Netezza

Netezza was one of such appliances launched in 2003 worlds' first data warehousing appliances, raising a business slogan we are different in terms performance, value and simplicity. There were many raised eyebrows in the industry as the competency of the existing big partner like Oracle, Microsoft, and Teradata. There are many questions by then the existence and future of the current data warehouse would there be an opportunity for Netezza Data Warehouse Appliance. Being cheaper and reduce nearly the 10% operational costs on yearly basis, it took the momentum over the existing RDBMS and MPP vendors. It adopted a faster momentum and performance more than 100x conventional data warehouses and as much as the half of the total cost of ownership. Netezza Data Warehouse Appliance provides a hardware architecture that is designed to execute a specific software system with overwhelming goal of data warehouse performance improvement.

Netezza is black-box solution, without separate components like database, server and storage. Netezza use a concept of distribution across the nodes SPU and field-programmable gate array hardware. Netezza DWA reduces effort of hardware tuning, database tuning and software tuning or upgrading hardware like adding SUN Fire 15K or 20k or add HP Superdome on the operating servers. Netezza also uniquely placed amongst the other vendors' right in the processing storage device and provide massively parallel architecture that is truly asymmetric massive parallel – more than 800 nodes in our larger systems. Netezza provided examples with many proof concept solution with various domain and industry vendors. Netezza manifested Google search engine appliance which allows an organization to quickly and easily provide a search engine capability for their own website or intranet simply by plugging in an appliance. Netezza DWA architecture is designed for query processing rather than transaction processing. Catalina Marketing is able to conduct shopping basket analysis against 70% of US point of sale data. Virgin media in the UK talk about their marketing campaigns are 3x more effective. New York stock market presented by IBM Netezza and Revolution R Analytics on Catastrophic events of stock market Flash Crash, the process runs the FINRA on TWINFIN node rules on historic data performed a back test calculation shown an improvement time to longest data in 2 hrs from a normal process of 8 to 10 hrs.

12.1.1 Features of Netezza Data Warehouse Appliance

- Data Modeling : Innovative, less usage of aggregate tables – without offline rebuilding, Standalone data structures
- No need to define the referential keys and corresponding look-up tables
- Less cumbersome with respect to defining the indexes, constraints and triggers
- Inbuilt database access of distribution of data pre-defined with specific patterns avoids full table scans
- Effective implementation of distribution of data vs partitioning
- ZoneMap acceleration for specific key on each table
- Unlike RDBMS, in DWA Hardware facilitates more I/O intensive operations like sorting, aggregations, in-line subqueries
- Interfaces to inbuilt database analytics support of SDK, IDE and Parallel Analytics Engines

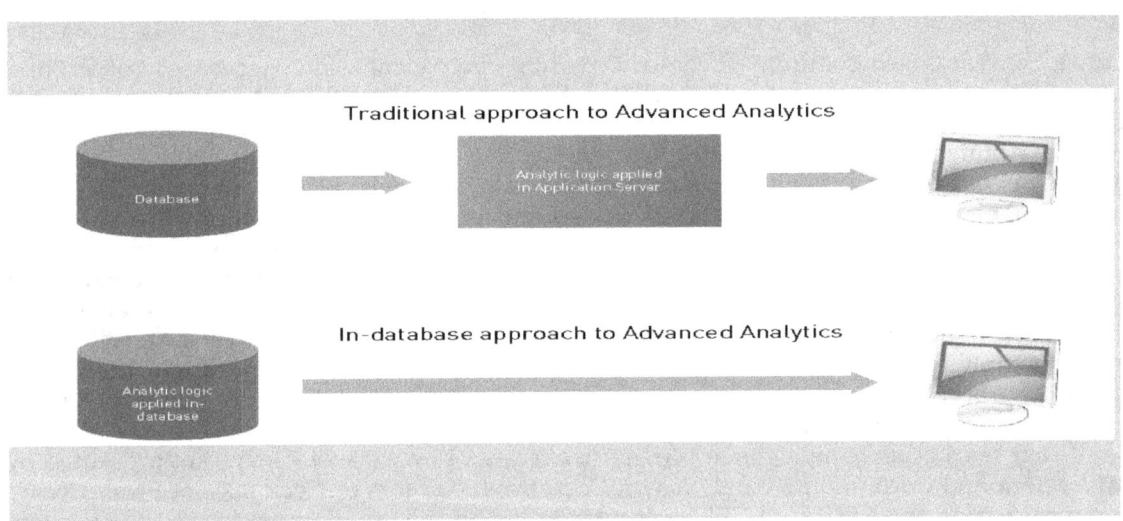

Figure 12.1.1 – Traditional Vs In-database Analytics

12.1.2 High Availability Architecture Implementation Using Netezza DWA

Netezza Data Warehouse Appliance on N1001 a Twin fin distributed replication block devices (DRBD) mode with auto replication and mirroring of data across the SPUs and is business intelligence specific to what data is available and what is known. Provide strong advanced analytics support discovering the patterns of data and processing information and framework with in-built database analytics with UDFS, external language interfacing with SDKs and IDEs. Netezza supports active replication across the nodes and host level configuration needs to be done to reflect the process flow.

Steps involved in implementing the High availability

- Replication - To maintain the availability of data at DB level.
- Recovery   - Back-up strategy – automatic flashback memory – restoration.
- Program level concurrent read and writes: facilitates availability for dirty reads – that is the records are not available to through a concurrent commit process; they have availability for a read.
- Read Consistency – Row level locking facilitating the   handling the concurrent transaction

- Serialization
-

12.1.3 Process Flow Diagram for Netezza HA

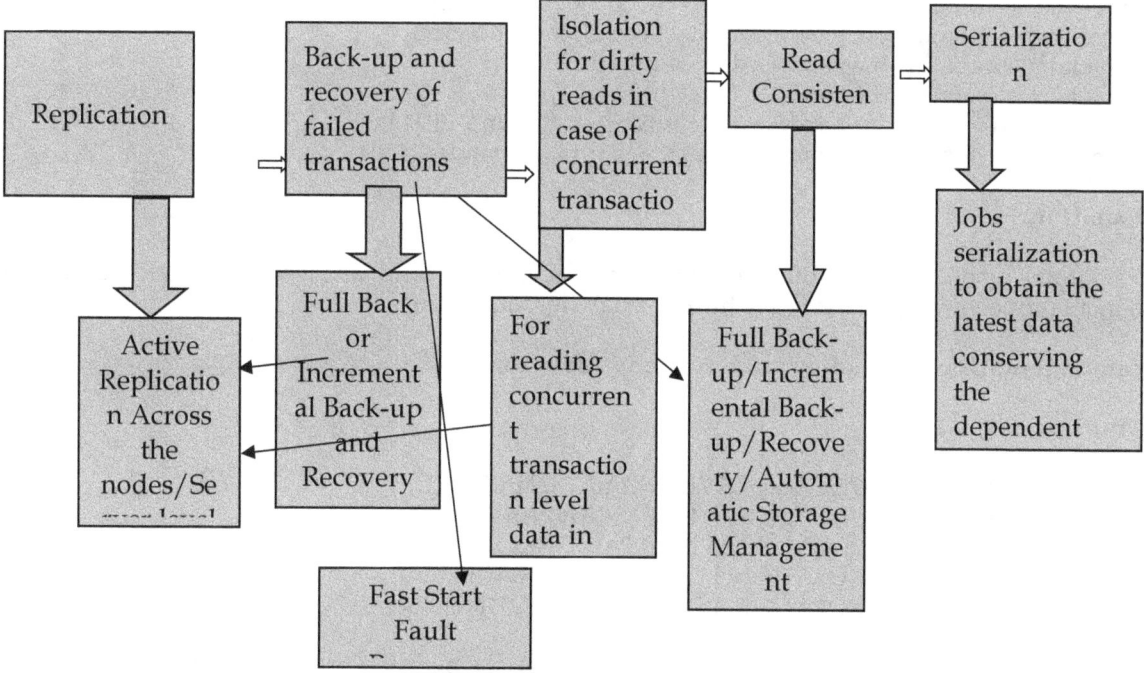

Figure 12.1.2 – Netezza Database High Availability Implementation

12.1.3 Netezza Replication

The Netezza high availability (HA) solution uses Linux-HA and Distributed Replicated Block Device (DRBD) as the foundation for cluster management and data mirroring. All the TwinFin models are HA systems, which mean that they have two host servers for managing Netezza operations. Distributed Replicated Block Device (DRBD) is a block device driver that mirrors the content of block devices (hard disks, partitions, logical volumes, and so on) between the hosts. Netezza uses the DRBD replication only on the /nz and /export/home partitions. As new data is written to the /nz partition and the /export/home partition on the primary host, the DRBD software automatically makes the same changes to the /nz and /export/home partition of the standby host.

Back-up and Recovery Strategy:
Nzback-up
nzbackup -db <db_name> - differential
nzbackup -db <db_name> - cumulative

NZRestore
nzrestore -db db1 -u user -pw password -dir /home/user/backups –
nzrestore -db new_db1 -sourcedb db1 -schema-only -u user –pw password-dir/home/user/backups
nzrestore -users -u user -pw password -dir /home/user/backups
nzrestore -db <dbname> -connector <filesystem> -dir <root dir>[<opts>] -tables <table1> [...]

NZReclaim

```
nzreclaim -u admin -pw password -blocks -db emp
nzreclaim -u admin -pw password -records -db emp
nzreclaim -u admin -pw password -startEndBlocks -db emp
nzreclaim -u admin -pw password -blocks -validate -db emp
nzreclaim -u admin -pw password -resume -db emp
nzreclaim -u admin -pw password -resume -t supplies
nzreclaim -u admin -pw password -scanBlocks -db emp-allTbls
nzreclaim -u admin -pw password -scanRecords -db emp-allTbls
```

Isolation

Facilitates availability of dirty reads – that is the records are not available through a concurrent commit process, they have availability for a read.

Implementation of Read-consistency and Isolation

Transaction level

```
SET TRANSACTION ISOLATION LEVEL READ COMMITTED;
SET TRANSACTION ISOLATION LEVEL SERIALIZABLE;
SET TRANSACTION ISOLATION LEVEL READ ONLY;
ALTER SESSION SET ISOLATION_LEVEL SERIALIZABLE;
ALTER SESSION SET ISOLATION_LEVEL READ COMMITTED;
```

These isolation levels prevent the following occurrences between concurrent transactions: Dirty reads — transaction reads data written by concurrent uncommitted transactions. Non repeatable reads — transaction re-reads data it previously read and finds that the data has been modified by another transaction (that committed since the initial read). Phantom read — A transaction re-executes a query returning a set of rows that satisfy a search condition and finds that the set of rows has changed due to another recently committed transaction.

Serialization

Multiversioning — each transaction sees a consistent state that is isolated from other transactions that have not been committed. Because of the Netezza architecture, the hardware can quickly provide the correct view to each transaction. Serialization dependency checking — concurrent executions that are not serializable are not permitted. If two concurrent transactions attempt to modify the same data, the system automatically rolls back the latest transaction. This is a form of optimistic concurrency control that is suitable for low-conflict environments.

12.1.4 Netezza SQL Functional Categories

Concurrent update or delete commands against different tables are permitted, with some restrictions that are needed to ensure serializability. For example: If transaction 1 selects from table A and updates (or deletes from) table B, while transaction 2 selects from table B and updates table A, Netezza SQL rolls back one or the other (typically the transaction that started more recently). This is called the cross-update case. If there is a cycle of three or more transactions (transaction 1 selects from A and updates B, transaction 2 selects from B and updates C, transaction 3 selects from C and updates A), the Netezza SQL rolls back one of the transactions in the cycle. Use the nzrecliam command to lock the table

SET THE READ ONLY OR READ WRITE AT SESSION LEVEL:
SET SESSION {READ ONLY | READ WRITE}

12.2 Oracle Exadata

Oracle Exadata as data warehouse as a strong competence with fast growing market of IBM Netezza. It layered and continued the customer base of Oracle RDBMS clients and vendors. Oracle's Optimized Warehouses offer pre-validated configurations and the database software comes pre-installed. In September 2008 Oracle began offering a more classic appliance offering, the HP Oracle Database Machine, a jointly developed and co-branded platform that Oracle sold and supported and HP built in configurations specifically for Oracle. In September 2009, Oracle released a second-generation Exadata system, based on their newly acquired Sun Microsystems hardware. Oracle Exadata is a database appliance with support for both OLTP (transactional) and OLAP (analytical) database systems. Oracle designed the database, operating system (based on the Oracle Linux distribution), and storage software whereas HP designed the hardware for it. After Oracle's acquisition of Sun Microsystems, in 2010 Oracle announced the Exadata Version 2 with improved performance and Sun storage systems.

Software caches database objects in flash memory, replacing slow, mechanical I/O operations to disk with rapid flash memory operations. Software also provides logging feature to speed database log I/O. Exadata storage cells determine which rows contain values that are being queried. Smart Scan only returns blocks that are relevant to the compute nodes. Storage cells can take over data intensive processing from compute nodes. Storage cells keep track on maximum and minimum values stored in different areas and use those values to determine where predicates cannot exist. This allows the storage cell to not have to read the area at all thus saving time and processing cycles.

12.2.1 Database servers X5.2/4.8/4.2

Each Database Server with
* 2 x Eighteen-Core Intel Xeon E5-2699 v3 Processors (2.3 GHz)
* 256 GB Memory (expandable to 768GB)
* Disk Controller HBA with 512MB Battery Backed Write Cache
* 4 x 600 GB 10,000 RPM Disks
* 2 x QDR (40Gbit/s) Ports
* 4 x 1/10 Gb Ethernet Ports (copper)
* 2 x 10 Gb Ethernet Ports (optical)
* 1 x ILOM Ethernet Port
* 2 x Redundant Hot-Swappable Power Supplies

The Exadata X5-2 introduces an elastic scale-out architecture in which additional database nodes and storage cells can be added. The configurations (Eighth, Quarter, Half and Full) still exist but are not the only configuration options.

12.2.2 Storage Servers X4-2/2.2

Each Storage Server has
* Processors: 2 x Six-Core Intel Xeon E5-2630 v2 (2.6 GHz) Processors
* Exadata Smart Flash: Cache 3.2 TB
* System Memory: 96 GB
* Disk Controller: Disk Controller HBA with 512MB Battery Backed Write Cache
* InfiniBand Connectivity: Dual-Port QDR (40Gbit/s) InfiniBand Host Channel Adapter
* Power Supplies: Dual-redundant, hot-swappable power supply
* Disk Drives:
** 12 x 1.2 TB 10,000 RPM High Performance or
** 12 x 4 TB 7,200 RPM High Capacity
** For raw disk capacity, 1 GB = 1 billion bytes. Actual formatted capacity is less.
* Remote Management: Integrated Lights Out Manager (ILOM) Ethernet port

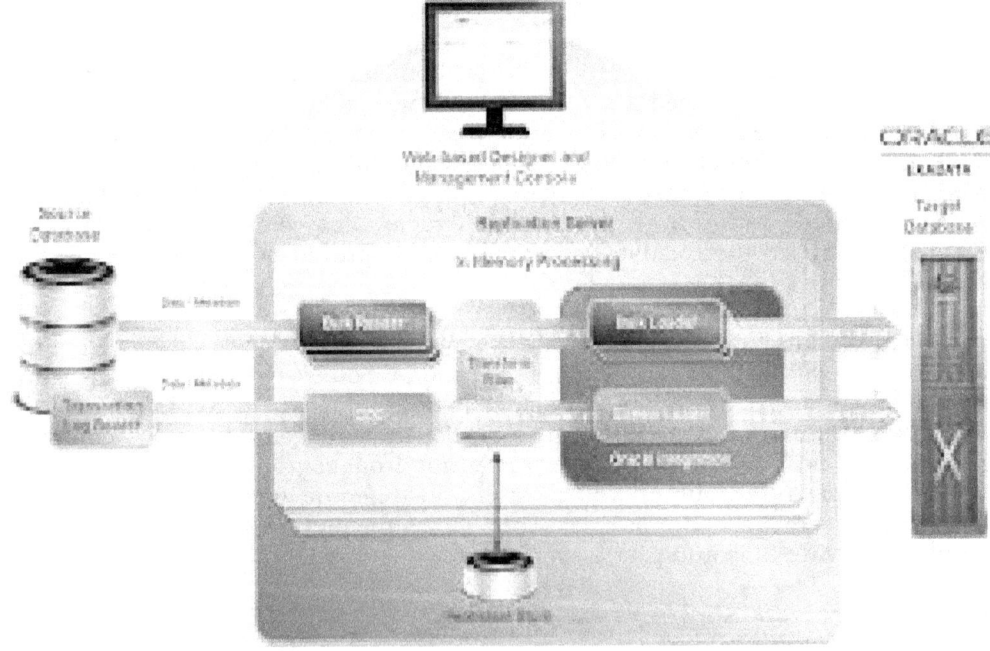

Figure 12.2.1 – Oracle Exadata Web based Designer and Management Console

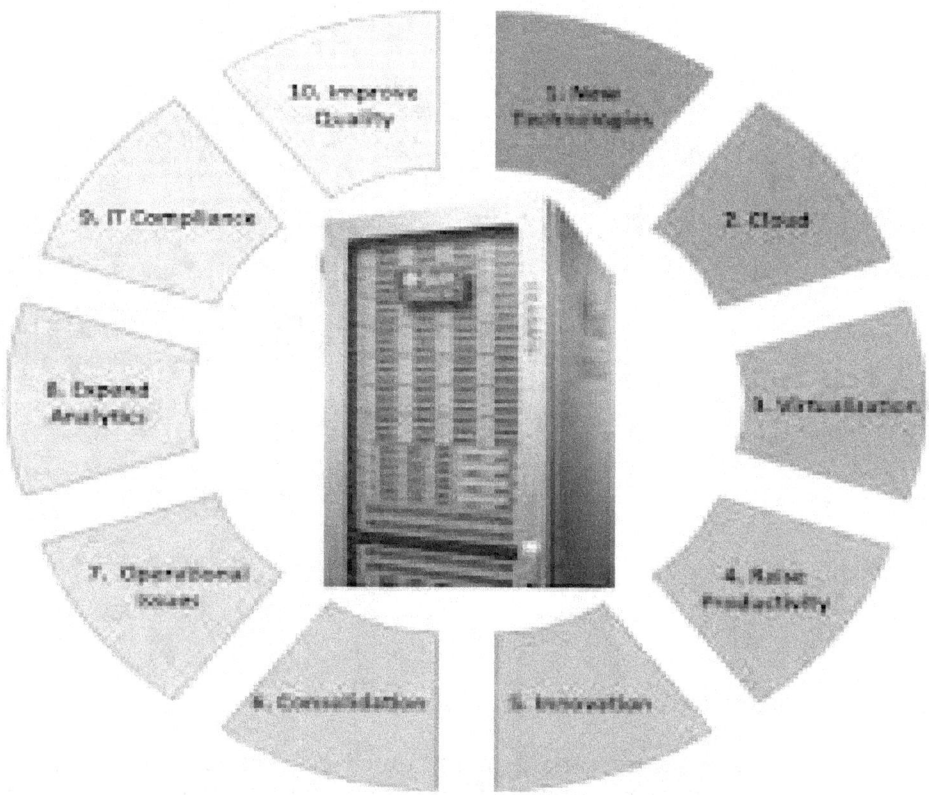

Figure 12.2.2 – Oracle Exadata Storage Server

12.3 Microsoft Data Allegro

Microsoft supported frame work of share everything architectures added new DATAllegro was a company that specialized in data warehousing appliances. DATAllegro from Microsoft architecture was implemented on commodity hardware from OEMs such as Dell Computer Corp., Cisco Systems Inc., and EMC Corp. DATAllegro - like Netezza - used open source software stack (Ingres DBMS running on Linux). Microsoft announced it had acquired DATAllegro as of September 2008. SQL Server Parallel Data Warehouse (PDW) is the successor product to DATAllegro on Windows Server using a version of the SQL Server database engine.

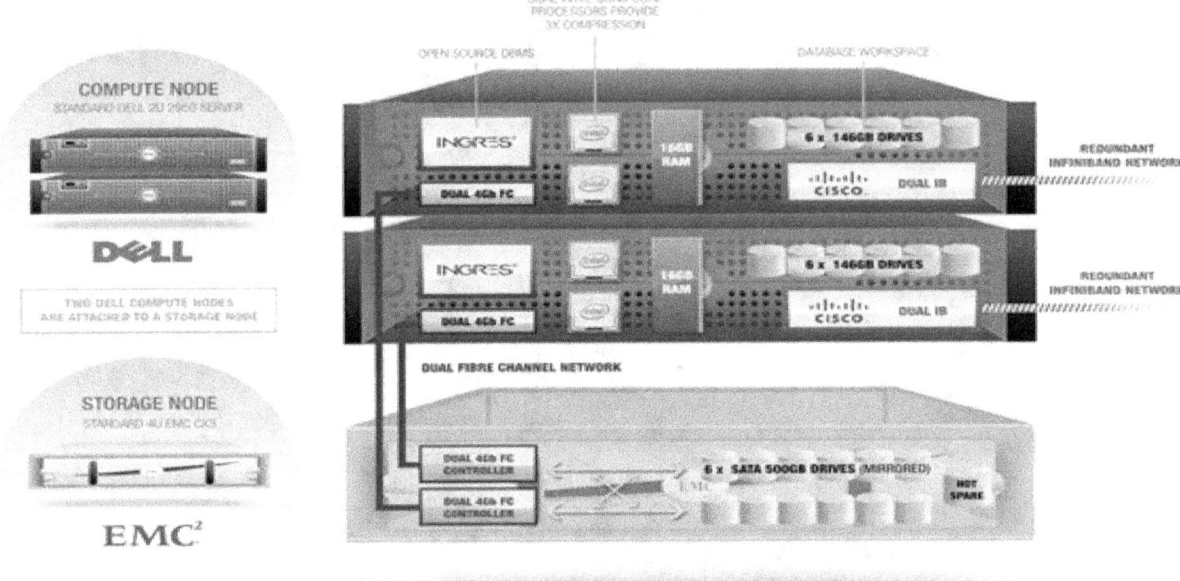

Figure 12.3.1 – Microsoft Data Allegro Server Architecture

Microsoft Data Allegro each server runs a highly tuned copy of the Ingres DBMS on SuSe Linux. Our proprietary software turns these separate databases into a massively parallel, shared nothing database system that offers incredibly good performance, especially under complex mixed workloads. The appliance model is key to getting great performance. Tuning a large database using traditional approaches is extremely difficult and requires highly skilled DBAs. One of the main problems is the difficulty of understanding and tuning the interface between the DBMS software and the underlying OS and hardware platform. Database vendors such as Oracle and Microsoft have to build their software to run on any hardware. Hence there are a plethora of tuning parameters and options for the DBA and sys admins to setup. In the appliance model, we have the luxury of controlling the entire software and hardware stack from SQL to storage. As a result, we can hide all of the complexity. Another very important aspect of performance is ensuring sequential reads under a complex workload. Traditional databases do not do a good job in this area - even though some of the management tools might tell you that they are! What we typically see is that the combination of RAID arrays and intervening storage infrastructure conspires to break even large reads by the database into very small reads against each disk. The end result is that most large DW installations have very large arrays of expensive, high-speed disks behind them - and still suffer from poor performance. Through a lot of trial and error, smart engineering and code changes to the database engine, we've been able to create a platform that sustains sequential reads - even under very high levels of concurrency. This allows us to use relatively low-cost, high-capacity SATA disk drives and therefore to provide a very high price/performance ratio.

12.4 Sun Microsystem GreenPlum

Greenplum was a big data analytics company Greenplum's products include its Unified Analytics Platform, Data Computing Appliance, Analytics Lab, Database, HD and Chorus. Greenplum was acquired by EMC Corporation in July 2010 and then became part of Pivotal Software in 2012. It was part sun micro systems in 2013. The Greenplum Database builds on the foundations of open source database PostgreSQL Greenplum Database's parallel query optimizer converts each query into a physical execution plan.

Greenplum's optimizer uses a cost-based algorithm to evaluate potential execution plans, takes a global view of execution across the computer cluster, and factors in the cost of moving data between nodes. The resulting query plans contain traditional relational database operations as well as parallel "motion" operations that describe when and how data should be transferred between nodes during query execution Commodity Gigabit Ethernet and 10 Gigabit Ethernet technology is used for the transfer between nodes. During execution of each node in the plan, multiple relational operations are processed by pipelining: the ability to begin a task before its predecessor task has completed, to increase effective parallelism. For example, while a table scan is taking place, rows selected can be pipelined into a join process. Internally, the Greenplum system utilizes log shipping and segment-level replication and provides automated failover. At the storage level, RAID techniques can mask disk failures. At the system level, Greenplum replicates segment and master data to other nodes to ensure that the loss of a machine will not impact the overall database availability. In 2009 technology was announced to use parallel streams of data for extract, transform and load operations. This technology is exposed to customers via a programmable "external table" interface and a traditional command-line loading interface.

In addition to traditional Structured Query Language (SQL), in 2008 support was announced for Map Reduce queries within a parallel dataflow engine, to run analytics against datasets stored in and outside of the Greenplum Database. For each table (or partition of a table), database administrators can select the storage, execution and compression settings that suit the way that table will be accessed. Greenplum DB transparently abstracts the details of any table or partition, allowing a variety of underlying models: traditional row-oriented tables, optimized for read-mostly scans and bulk append loads, or column-oriented.[20] Database administrators also can tune the storage types and compression settings of different partitions within the same table. Greenplum HD is a supported version of Apache Hadoop. It includes Hadoop's Distributed File System (HDFS), Hive, Pig, HBase, and ZooKeeper Greenplum Chorus is a social network portal for data science teams. The Greeplum Data Computing Appliance (DCA) is a physical computer appliance to integrate structured data, unstructured data, and partner applications such as business intelligence. A special version of DCA integrated with SAS software was released in April 2011. Greenplum Command Center software displays interactive dashboards to collect performance metrics and manage system health for Greenplum products. Monitored data is also stored for historical reporting. Greenplum Analytics Lab was a data science consultation service, renamed Pivotal Data Labs in 2013.

Greenplum Database was supported for production use on SUSE Linux Enterprise Server 10.2 (64 bit), Red Hat Enterprise Linux 5.x (64-bit), CentOS Linux 5.x (64-bit) and Sun Solaris 10U5+ (64-bit). Greenplum Database was supported on server hardware from a range of vendors including HP, Dell, Sun and IBM Greenplum Database was supported for non-production (development and evaluation) use on Mac OS X 10.5, Red Hat Enterprise Linux 5.2 or higher (32-bit) and CentOS Linux 5.2 or higher (32-bit). Greenplum had customers in vertical markets from financial services, telecommunications, Internet, retail, transportation and pharmaceuticals industries They included Silver Spring Networks, Zions Bancorporation, Reliance Communications, NYSE Euronext, Orbitz, Havas Digital, China Unicom, and Tagged.

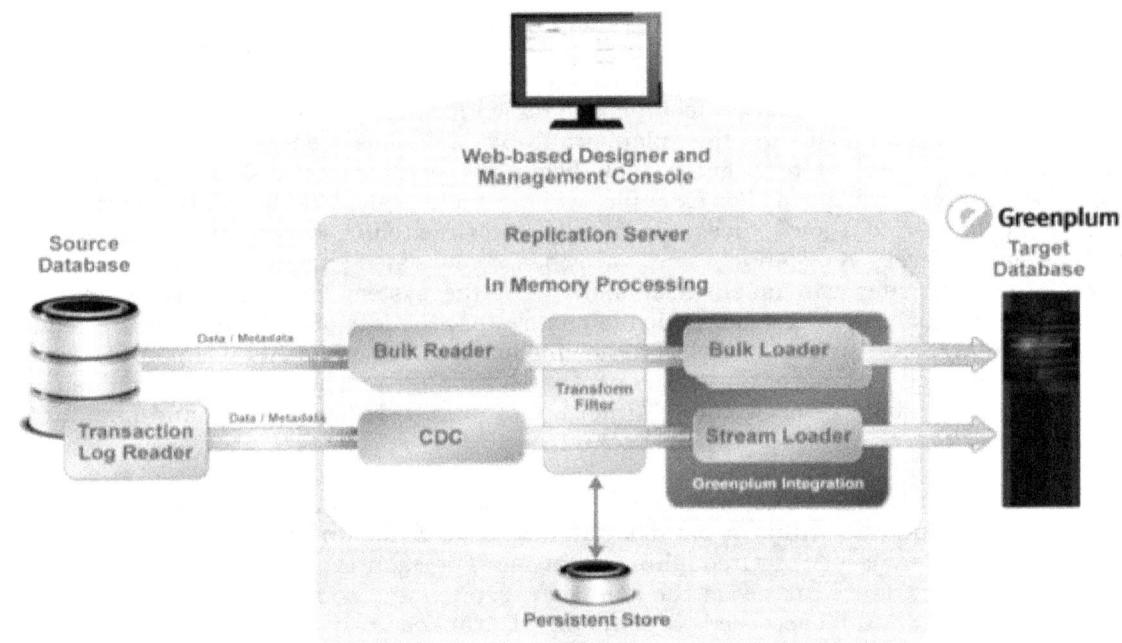

Figure 12.4.1 – GreenPlum Web Based Designer and Management Console

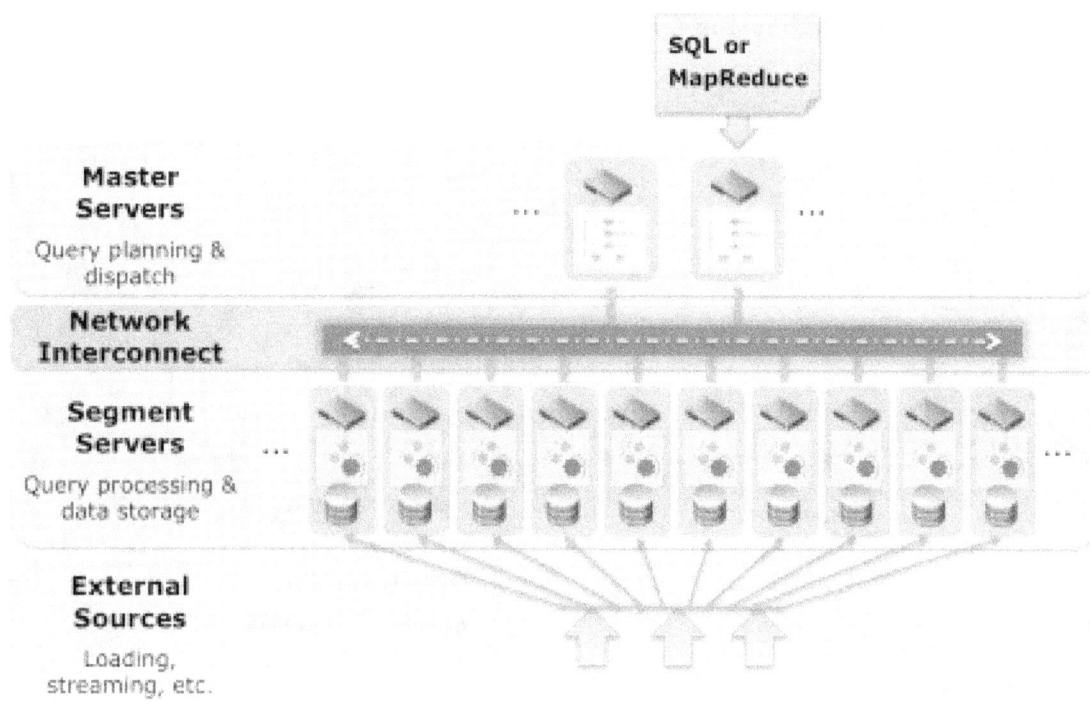

Figure 12.4.2 – GreenPlum SQL Map Reduce Framework

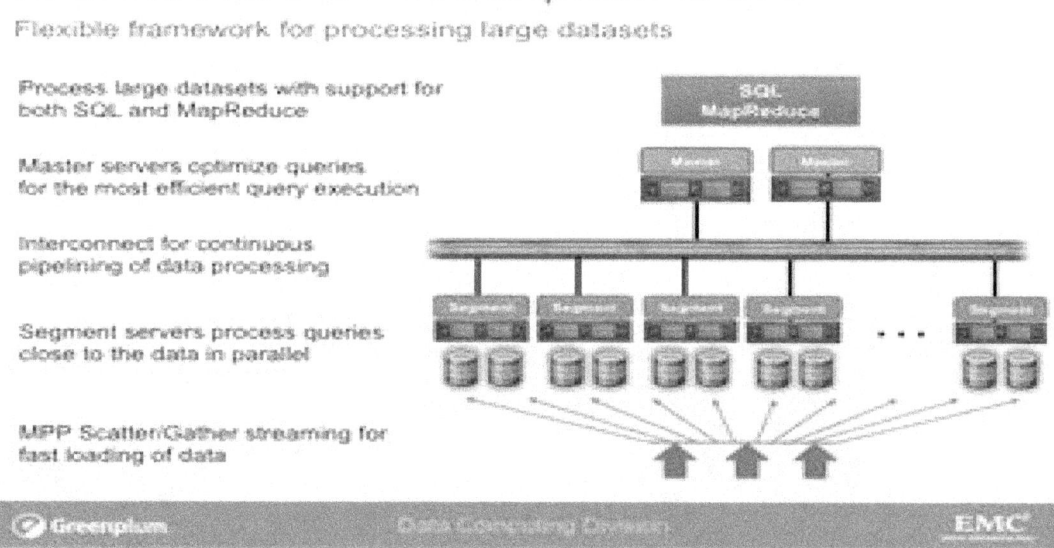

Figure 12.4.3- Architecture for GreenPlum DCA (Data Computing Appliance)

12.5 Comparing Data Warehousing Appliances

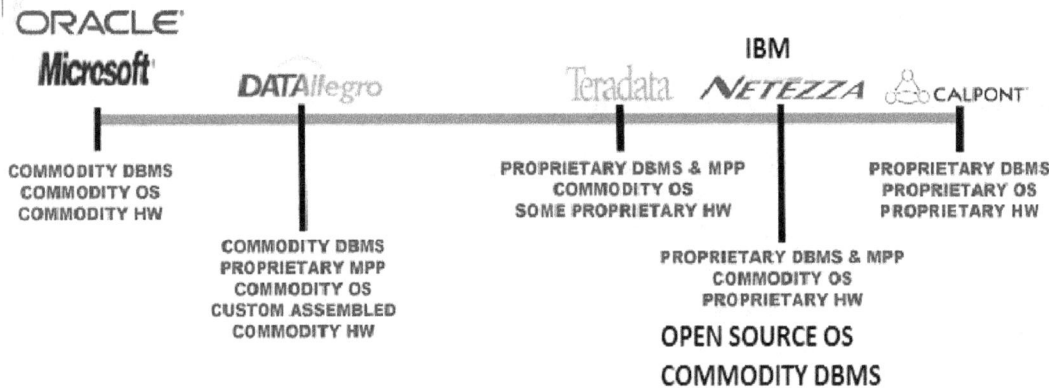

Figure 12.5.1 – Comparing Data Warehouse Appliance

CHAPTER 13
ADVANCED ANALYTICS

The combination of an increasingly complex world, the vast proliferation of data, created demand using analytics within organizations. Advanced Analytics models improve operations tackle and solve your most complex problems and create unique predictive capabilities for intelligent decisions. The advanced innovative research in data analytics and optimization helps to establish focus based strategies, to enhance performance, to improve operational efficiency and to better understand how organizations are applying analytics today, prioritizing their future investments, and transforming insights into action. There is a major difference between advanced analytics and business intelligence. Advanced Analytics basically look through discover and understand the patterns of behavior and activity through deep data exploration and ad-hoc analysis. Business Intelligence on the other hand presents what the data is available and what is known. In-place analytics, processing will mean that network bottlenecks can be eliminated, while the high performance processing is renowned delivers significantly faster time to insight. Business Intelligence as such fewer limitations and challenges with respect to ease of use and flexibility in reformatting the information, processing it on a thin layer or web based. Advanced Analytics thus eliminates any such bottlenecks as being a platform built on robust runtime environment. Most of the processing and handling of information happens on the hardware. Data Warehouse Appliance solutions being unique platform in establishing such framework, where the complex information is handled in much more efficient way. It is also cost effective in terms of reducing post implementation support. Advanced analytics refers to future-oriented analyses that can be used to help drive changes and improvements in business practices.

13.1 MapReduce

One of such system for distributing code across multiple processors is MapReduce. The system allows Map functions and Reduce framework together to combine the results from code multiple servers parallel. The map functions allow data to be organized into key value pair list. The reduce functions processes the lists in parallel to produce a set values that are desired result of the initial query. For instance the Reduce functions might return number of electronic products sold in a region during a specific quarter. This can be returned from each node, resulting in less data to compile to determine the final result. Google has used this framework to process petabytes of unstructured data it mines for service.

13.2 Advanced Analytics with Apache Spark

Apache spark an open source, has as its architectural foundation the resilient distributed dataset (RDD), a read-only multiset of data items distributed over a cluster of machines, supports Dataset API. Spark and its RDDs were developed in 2012 in response to limitations in the MapReduce cluster computing paradigm, Spark facilitates the implementation of both iterative algorithms, that visit their data set multiple times in a loop, and interactive/exploratory data analysis, i.e., the repeated database-style querying of data. Apache Spark requires a cluster manager and a distributed storage system. For cluster management, Spark supports standalone (native Spark cluster), Hadoop YARN.

13.2 Advanced Analytics with Hadoop Scala

Hadoop Scala is vendor managed supports the following:

- Adjust automatic messaging to reflect customer habits and trends based on factors such as price change, weather, environmental factors, time, day of the week and season
- Align inventory, specific business objective and historical data maximize profits
- Highlight messages that reflect what potential buyers want ot need to see, in addition to what products or information you want market most.
- Use automated feedback for systems to optimize your messages and to maximize sales.

13.3 Netezza Appliance for Pure data Analytics

Data warehouse and analytic appliance with strong MPP parallel processing data structures. Support API, access methods uses the proprietary protocols for HTTP API and SOAP-Based API, JDBC, ODBC driver support.

CHAPTER 14
CLOUD COMPUTING- SOFTWARE AS SERVICE

The past decade has been rapid evolution of the business landscape and business organizations are increasingly realizing the need for more scalable and flexible information technology architecture. The reducing borders of burden of regulation and compliance is further amplifying business expectations, for small vendors and middle scale business partners. The emergence of IT industry has been quick in its response with the innovation in technologies such as virtualization, SOA, Web 2.0, which is base for the development of cloud platforms. The cloud model thus enabled by SOA provides flexibility and scalability using external computing and processing power in the form of real time e-services. The primary benefits driven by this model are business agility with lower costs, enabling organizations to respond quickly and effectively to the ever changing business environment apart from support for robust run-time environment. As per IT pundits, is one of the future evolutions of computing and investment IT industry? But, regulatory and other major challenges prevent the cloud implementations.

The seven basic principles of cloud defining to a structured methodology of Process, Procure and Implement change management model, having the support virtualization model and off-site on demand acquisition. Many of the process adoptions can be very ad-hoc, none of the requirements needs to be very well formulated, like in many traditional platforms. For example user level report needs to be caned in a batch processing, can be processed on demand thus provide lot of outsourcing of non-platform oriented shared services. Web-based reporting supports BO, Cognos, OBIEE and Micro Strategy etc., all do have the same functionality, but on the same vendor platforms and nodes. Azure, Google, 1010 are other cloud models.

The cloud model facilitates:
- Utility model of computing
- Lower entry and maintenance costs
- Massive processing power on-demand
- Faster time to develop applications
- Multiple ways to consume the BI capability
- Multiple deployment and delivery models

14.1 Types of Clouds

Private Cloud, Public Cloud, Hybrid Cloud

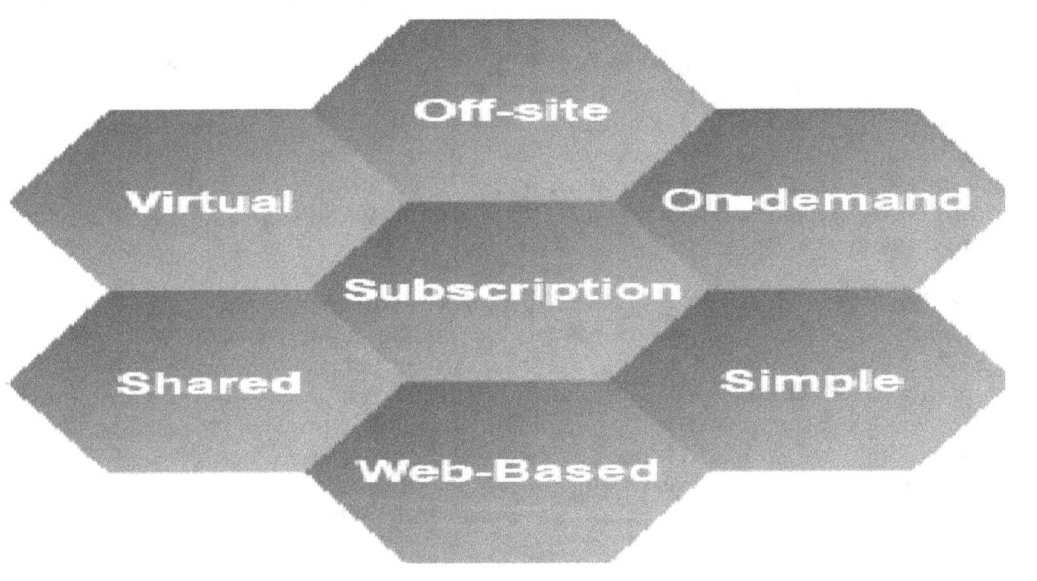

Figure 14.1.1 - Seven Principles of Cloud Computin

14.2 Cloud Architecture and Web Services

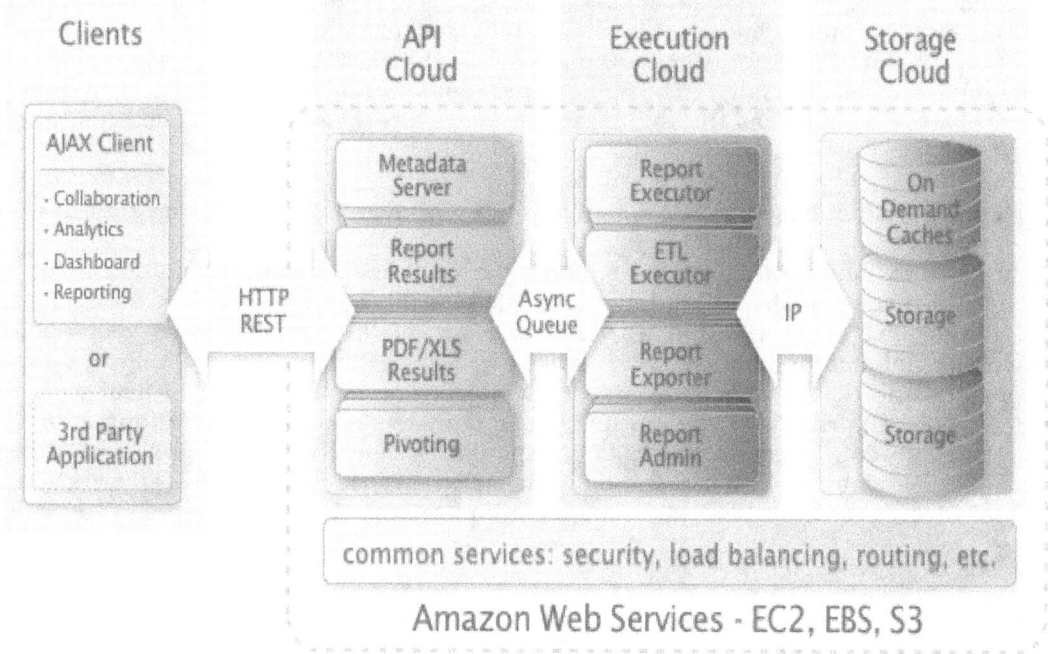

Figure 14.2.1 – Amazon Web Services Cloud Environment

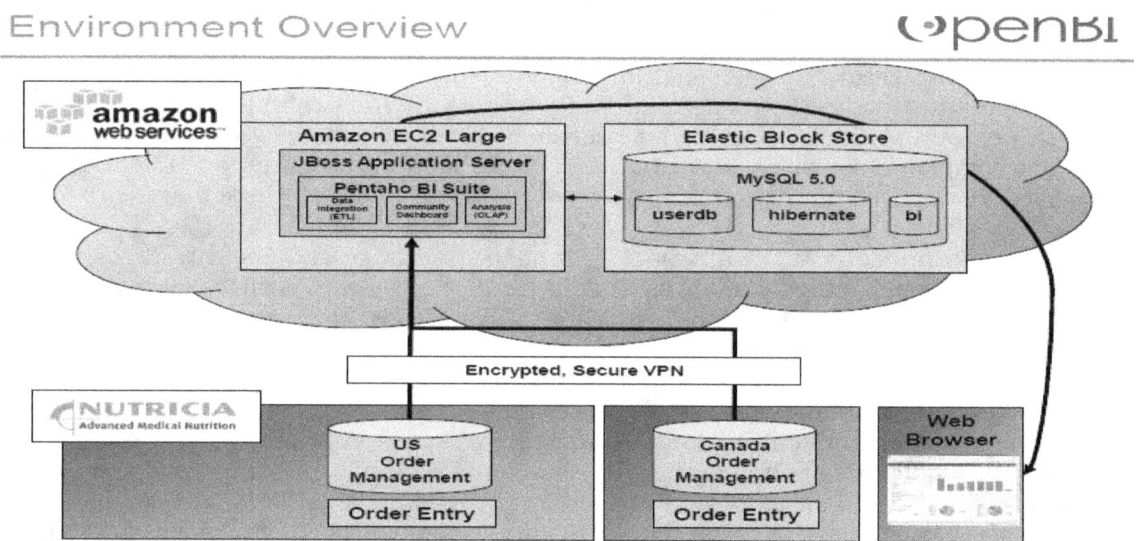

Figure 14.2.2 – Open BI Cloud Environment Overview

14.3 Cloud limitations and challenges

The operational costs may be challenging issues items of external vendor service cost payment models. The other challenges of data quality, security and privacy of external data sourcing may be limitation for the banking and financial, telecom, manufacturing and healthcare and pharma domains. The next generation prescription processing system of health and pharma domain has the web-interface, not addressing the above said limitations and challenges.

14.3.1 Use of the cloud-based services will be bound by legal and regulatory policies

Many policies apply when sensitive data such as personal or financial information is moved outside the jurisdiction of those regulations. While cloud-based services would largely not promote the keeping of data within an enterprise's national borders, some geographical requirements or national preferences will restrict where data can legally reside. For example, a U.S. company that keeps personal data on European customers needs to deal with the European Union (EU) privacy laws, even if that company does not have operations in an EU country. However, data location restrictions should not be forced on providers unless compliance is mandatory. Restrictions on data location will reduce the efficiency and flexibility of data architectures, and could minimize the business benefits of cloud computing.

14.3.2 Case Study on cloud under various Regulatory acts

U.S. and EU Privacy and Data Security Law Issues

Below is a summary of some key U.S. and EU legal and regulatory considerations that may come into play in the cloud computing context.

- Service Provider Restrictions- Certain U.S. regulatory frameworks require data owners to ensure that their third party service providers are capable of maintaining the privacy and security of personal information entrusted to them. Two federal privacy laws that restrict the activities of service providers are the Health Insurance Portability and Accountability While these requirements do not restrict the geographic movement of a company's personal information (unlike the laws in the European Union), they do place restrictions on the use of service providers regardless of where they, or the data, are located
- HIPAA Restrictions on Health Data - Through its Privacy and Security Rules, HIPAA imposes significant restrictions on the disclosure of protected health information
- Gramm-Leach-Bliley Act - For entities subject to GLB, the use of a cloud provider would be subject to similar restrictions. GLB's Privacy and Safeguards Rules restrict financial institutions from disclosing consumers' nonpublic personal information to non-affiliated third parties
- State Information Security Laws- A numbers of states impose a general information security standard on businesses that maintain personal information. These states, which include Arkansas, California, Connecticut, Maryland, Nevada, Oregon, Rhode Island, Texas and Utah, have laws requiring companies to implement reasonable information security measures
- State Breach Notification Laws- The use of cloud computing may raise concerns with respect to U.S. state breach notification requirements. Over 45 U.S. states and other jurisdictions have data security breach notification laws that require data owners to notify individuals whose computerized personal information has been subject to unauthorized access or acquisition
- Breach Provisions Under HITECH Act- The Health Information Technology for Economic and Clinical Health Act, Pub. L. 111-5, 123 Stat. 258-263, established new information security breach notification requirements that apply to a wide range of businesses that handle PHI and other health data

- European Union Regulatory Issues - Data protection authorities in the European Union recently have paid particular attention to cloud computing, largely in response to inquiries from vendors and prospective users of cloud technology seeking to ensure compliance with EU data protection requirements
- International Data Transfers - The restrictions placed on the international transfer of personal data by EU Member States raise particularly troublesome jurisdictional issues in the context of cloud computing

CHAPTER15
SELF SERVICE BUSINESS INTELLIGENCE AND MOBILE BI

Self-service BI allows your employees instant access to the business intelligence they need to operate at highest levels. Not only does this put your staff in the driver's seat for making faster decisions, but it also frees up your IT workers so they may focus on technology rather than data collection. Self Service facilitates the data in the hands of end-users is the ultimate driver of self-service BI. In other words, your employees are the ones who need the information, so why not let them decide which dashboards will best help them make informed decisions. In the old days, getting the BI you needed consisted of providing an explanation to data analysts only to find it would have worked better had you done this or that differently, technology can now adapt to the needs of the staffer.

15.1 Deploying Self Service BI Solution

How do you deploy self-service BI? First, assess your people, processes, technology and goals. After doing so, you should identify the business units across your organization, who access information from the current BI application. These people will likely be your early adopters when it comes to learning the new technology. In other words, choose the right software, and then begin the gradual process of deployment. How do you assimilate business users to the system? It starts with identifying your power users. Power users are the people who more quickly grasp self-service BI. In the beginning, you should work to educate your employees on the basics. Give them some pre-built dashboards to play with, and encourage experimentation. Show the ones who have an interest - those who aren't afraid to play with technology - how to create and modify existing dashboards. As their experience grows, you can not only teach them how to draw their own dashboards from scratch, but you can also benefit from their knowledge by helping them train new associates and those who are struggling with the new application.

What comes after assimilation? About 80 percent of all data requested by business users today can be standardized with the other 20 percent available through adjustments on exploratory dashboards. What that means for the employees and for the organization IT professionals is that more time can be spent on the functions that matter. On the employee side, decisions can be made with greater reliance on facts and therefore confidence. On the IT side, efforts can go more toward maintaining, troubleshooting, and strengthening the system. Self Service BI evolve to agile BI in which on the job functionality leads to continued efficiency gains associated with the maintenance and access to business intelligence. In the latest advancements Self Service BI achieved a greatest level of business management. With the Self Service BI pertinent data falls into the hands of people who need it the most can use it best-your employees.

15.2 Self Service BI Architecture

- End to end changeability of a BI solution by BI power user:
- Continual changes due nature of business being supported
- Changes impacting complete solution due to snowballing effect of changes on any part of the architecture
- Significant time required to implement solution change

The self-serviced BI platform has the following base components. There is a metadata backbone, followed by a dynamic ETL framework and the reporting framework which will utilize the metadata backbone. The extensible relational database structures support these frameworks. The metadata backbone contains the complete information regarding how the application structured. For example, a report has the information about the UML in the reporting layer; it will contain the details of the reports such as the number, name and type of fields in a report. The conventional adoption to the software development life cycle and self-service will reduce of cost re-design the platform repeatedly also to the process for deploying the solution, as the self-service BI uses the backbone of metadata and in-built ETL framework.

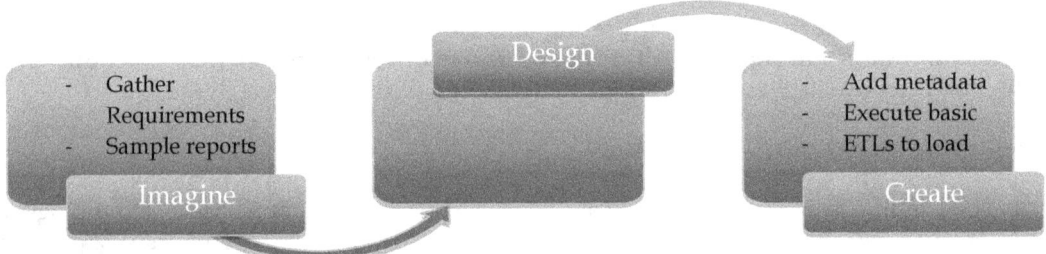

Figure 15.1 – Typical SDLC with Self-Service BI

15.3 Self Service BI ETL Framework

- End to end changeability of a BI solution by BI power user
- Continual changes due nature of business being supported
- Changes impacting complete solution due to snowballing effect of changes on any part of the architecture

15.4 Self Service BI Implementation

- Primary Components
- Metadata Backbone
- Dynamic ETL framework
- Dynamic Reporting framework
- Extensible relational database structures
- Metadata base contains
- Complete data flow information for a dynamic ETL based data flow system
- Reporting information such as data column mapping, formatting, calculations for the dynamic reporting framework
- Extensible relational database structures able to absorb addition of new database objects/schema

The ETL framework described supports metadata backbone. It uses the metadata base to get complete data flow information such as which tables to use for the extraction of the data. As a new generation, if you talk about using the Microsoft Technologies for implementation, SQL server integration services used to adapt and implement the ETL framework and. MS SQL server reporting services can be used to build the reporting framework.

15.5 Comparison of conventional BI life Cycle vs Self Serviced BI

In conventional BI SDLC cycle you will have 4 stages for implementing a data warehouse solution. In the self-service BI you will see only stages such as imagine, design and create. This would reduce of the cost and time that is required to collect the information from source systems and also benefit quick process implementation building an underlying framework for both ETL extraction and reporting. Other benefits would be to reduce the data cleaning process, also to follow to quick change management process. The primary focus and benefit of such a solution architecture would be easy and quick absorption of changes to solution. Thus, if we have a new business requirement which requires a change to the solution, all we have to do is add/modify or delete records in the metadata base. Since all other components rely on the metadata base for their activities these changes would result in the change being implemented throughout the solution. Thus, facilitates self-service BI architecture automates the change management of the solution.

CHAPTER 16
CLOUD WITH ETL INTERFACES, BIGDATA, DATA SCIENCE, DATA LAKES

16.1 Cloud with ETL Interfaces

The more software systems that we deploy to cloud environments, the greater the need will be to have an efficient integration strategy. Integration through messaging is possible through something like an on-premises integration server, or via a variety of cloud tools such as queues hosted in AWS or something like the Windows Azure Service Bus Relay. However, what if you want to do some bulk data movement with Extract-Transform-Load (ETL) tools that cater to cloud solutions? One of the market leaders in the overall ETL market, Informatica, has also established a strong integration-as-a-service offering with its Informatica Cloud. They recently announced support for Dynamics CRM Online as a source/destination for ETL operations.

Informatica Cloud supports a variety of sources/destinations for ETL operations and leverages a machine agent ("Cloud Secure Agent") for securely connecting on-premises environments to cloud environments. Instead of installing any client development tools, I can design my ETL process entirely through their hosted web application. When the ETL process executes, the Cloud Secure Agent retrieves the ETL details from the cloud and runs the task. There is no need to install or maintain a full server product for hosting and running these tasks. The Informatica Cloud doesn't actually store any transactional data itself, and acts solely as a pass through that executes the package (through the Cloud Secure Agent) and moves data around.

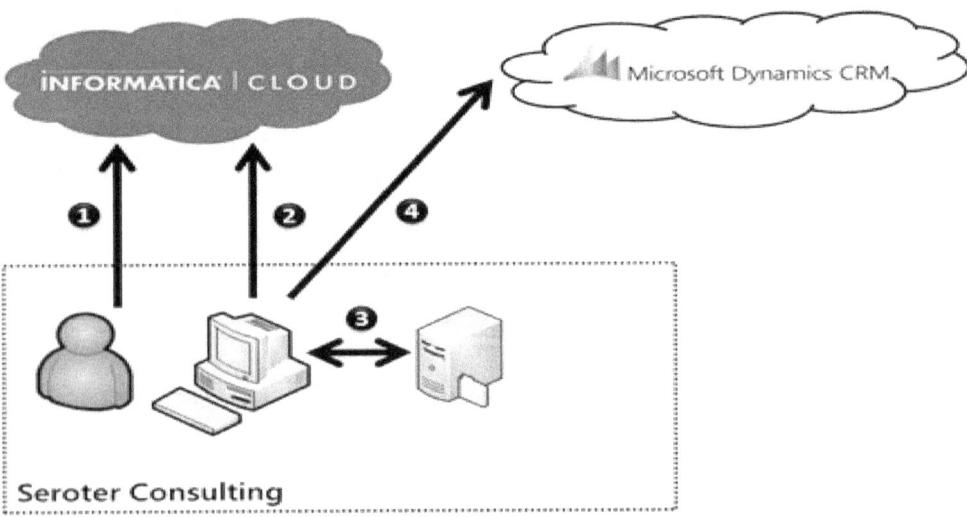

Figure 16.1.1 – Informatica Cloud

16.2 Informatica ETL Interface

The Informatica Cloud to define an ETL that takes a flat file from my local machine and copies the data to Dynamics CRM Online, the Secure Cloud Agent will communicate with the Informatica Cloud to get the ETL details, the Secure Cloud Agent retrieves the flat file from my local machine, and finally the package runs and data is loaded into Dynamics CRM Online.

16.3 Building the ETL Package

To get started, log-in to Informatica Cloud account and walked through their Data Synchronization wizard. In the first step, Name the 1st ask and chose to do an Insert operatio

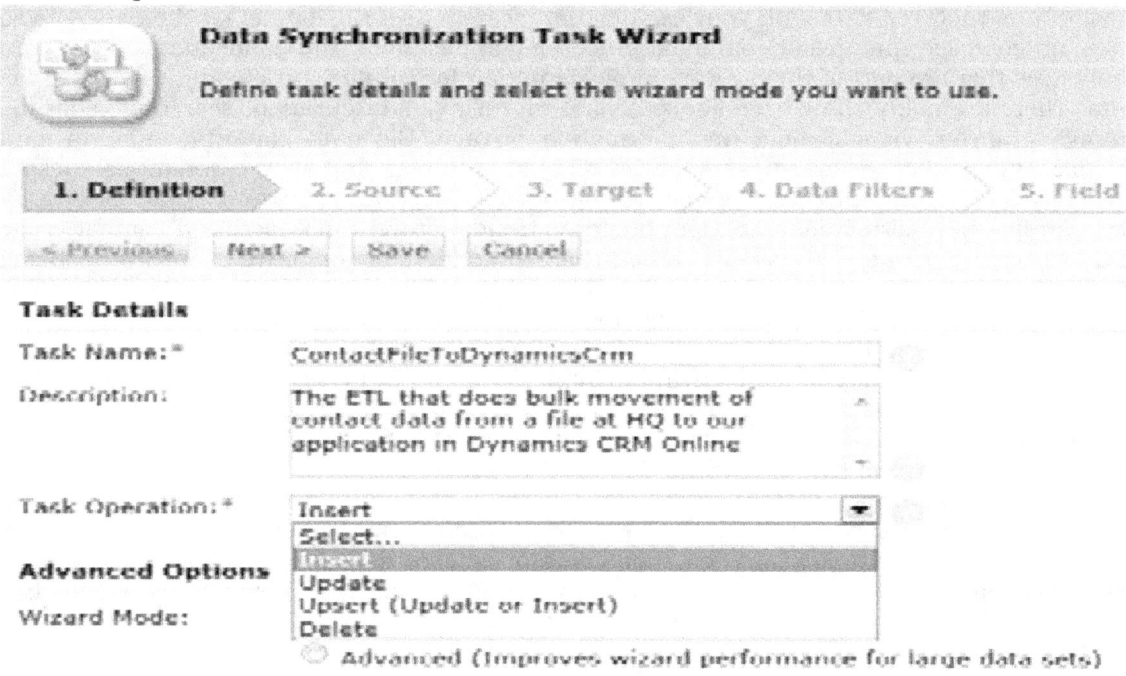

Next, chose to create a "flat file" connection type. This requires my Agent to have permissions on my file system, so set the Agent's Windows Service to run as a trusted account on my machine. With the connection defined, then choose to use a comma delimited formatter, and chose the text file in the "temp" directory.

16.3 Oracle service Cloud

This will enable you to build cloud with identity management.

16.5 BIGDATA

What is Bigdata?

Data Warehousing in the Age of the Big Data will help you and your organization make the most of unstructured data with your existing data warehouse. Big data is for the data sets so large or complex that traditional data processing applications are inadequate. There are challenges that are part of the big data implementation for volume, variety and velocity of the data. There are many challenges like analysis, capturing, data curation, searching, sharing, storage, transfer, visualization, and information privacy. Big data platform supports data science and predictive analytics. Analysis of data sets can find new correlations to spot business trends, prevent diseases and combat fraud detection and so on. Analysis of data sets can find new correlations, to "spot business trends, prevent diseases, and combat crime and so on. Scientists, business executives, practitioners of media and advertising and governments alike regularly meet difficulties with large data sets in areas including Internet search, finance and business informatics. Scientists encounter limitations in e-Science work, including meteorology, genomics, connect omics, complex physics simulations, and biological and environmental research. Data sets grow in size in part because they are increasingly being gathered by cheap and numerous information-sensing mobile devices, aerial (remote sensing), software logs, cameras, microphones, radio-frequency identification (RFID) readers, and wireless sensor networks.
Relational database management systems and desktop statistics and visualization packages often have difficulty handling big data. The work instead requires "massively parallel software running on tens, hundreds, or even thousands of servers". What is considered "big data" varies depending on the capabilities of the users and their tools, and expanding capabilities make Big Data a moving target. Thus, what is considered to be "Big" in one year will become ordinary in later years. "For some organizations, facing hundreds of gigabytes of data for the first time may trigger a need to reconsider data management options. For others, it may take tens or hundreds of terabytes before data size becomes a significant consideration.

16.6 Definition of Bigdata

Big data usually includes data sets with sizes beyond the ability of commonly used software tools to capture, curate, manage, and process data within a tolerable elapsed time. Big data "size" is a constantly moving target, as of 2012 ranging from a few dozen terabytes to many petabytes of data. Big data is a set of techniques and technologies that require new forms of integration to uncover large hidden values from large datasets that are diverse, complex, and of a massive scale. In a 2001 research report and related lectures, META Group (now Gartner) analyst Doug Laney defined data growth challenges and opportunities as being three-dimensional, i.e. increasing volume (amount of data), velocity (speed of data in and out), and variety (range of data types and sources). Gartner, and now much of the industry, continue to use this "3Vs" model for describing big data. In 2012, Gartner updated its definition as follows: "Big data is high volume, high velocity, and/or high variety information assets that require new forms of processing to enable enhanced decision making, insight discovery and process optimization. Additionally, a new V "Veracity" is added by some organizations to describe it.

16.7 Characteristics of Bigdata

Big data can be described by the following characteristics:

Volume – The quantity of data that is generated is very important in this context. It is the size of the data which determines the value and potential of the data under consideration and whether it can actually be considered Big Data or not. The name 'Big Data' itself contains a term which is related to size and hence the characteristic.

Variety - The next aspect of Big Data is its variety. This means that the category to which Big Data belongs to is also an essential fact that needs to be known by the data analysts. This helps the people, who are closely analyzing the data and are associated with it, to effectively use the data to their advantage and thus upholding the importance of the Big Data.

Velocity - The term 'velocity' in the context refers to the speed of generation of data or how fast the data is generated and processed to meet the demands and the challenges which lie ahead in the path of growth and development.

Variability - This is a factor which can be a problem for those who analyse the data. This refers to the inconsistency which can be shown by the data at times, thus hampering the process of being able to handle and manage the data effectively.

Veracity - The quality of the data being captured can vary greatly. Accuracy of analysis depends on the veracity of the source data.

Complexity - Data management can become a very complex process, especially when large volumes of data come from multiple sources. These data need to be linked, connected and correlated in order to be able to grasp the information that is supposed to be conveyed by these data. This situation, is therefore, termed as the 'complexity' of Big Data.

16.8 Architectures of Bigdata

In 2000, Seisint Inc. developed C++ based distributed file sharing framework for data storage and querying. Structured, semi-structured and/or unstructured data is stored and distributed across multiple servers. Querying of data is done by modified C++ called ECL which uses apply scheme on read method to create structure of stored data during time of query. In 2004 LexisNexis acquired Seisint Inc and 2008 acquired ChoicePoint, Inc and nd their high speed parallel processing platform. The two platforms were merged into HPCC Systems and in 2011 was open sourced under Apache v2.0 License. Currently HPCC and Quantcast File System are the only publicly available platforms capable of analyzing multiple exabytes of data.

In 2004, Google published a paper on a process called MapReduce. The MapReduce framework provides a parallel processing model and associated implementation to process huge amounts of data. With MapReduce, queries are split and distributed across parallel nodes and processed in parallel (the Map step). The results are then gathered and delivered (the Reduce step). The framework was very successful, so others wanted to replicate the algorithm. Therefore, an implementation of the MapReduce framework was adopted by an Apache open source project named Hadoop. MIKE2.0 is an open approach to information management that acknowledges the need for revisions due to big data implications in an article titled "Big Data Solution Offering. The methodology addresses handling big data in terms of useful permutations of data sources, complexity in interrelationships, and difficulty in deleting (or modifying) individual records. Recent studies show that the use of a multiple layer architecture is an option for dealing with big data. The Distributed Parallel architecture distributes data across multiple processing units and parallel processing units provide data much faster, by improving processing speeds. This type of architecture inserts data into a parallel DBMS, which implements the use of MapReduce and Hadoop frameworks. This type of framework looks to make the processing power transparent to the end user by using a front end application server. Big Data Analytics for Manufacturing Applications can be based on a 5C architecture (connection, conversion, cyber, cognition, and configuration). Big Data Lake - With the changing face of business and IT sector, capturing and storage of data has emerged into a sophisticated system. The big data lake allows an organization to shift its focus from centralized control to a shared model to respond to the changing dynamics of information management. This enables quick segregation of data into the data lake thereby reducing the overhead time.

16.9 Big data Technologies

Big data requires exceptional technologies to efficiently process large quantities of data within tolerable elapsed times. A 2011 McKinsey report suggests suitable technologies include A/B testing, crowdsourcing, data fusion and integration, genetic algorithms, machine learning, natural language processing, signal processing, simulation, time series analysis and visualization. Multidimensional big data can also be represented as tensors, which can be more efficiently handled by tensor-based computation, such as multilinear subspace learning.[39] Additional technologies being applied to big data include massively parallel-processing (MPP) databases, search-based applications, data mining, distributed file systems, distributed databases, cloud based infrastructure (applications, storage and computing resources) and the Internet. Some but not all MPP relational databases have the ability to store and manage petabytes of data. Implicit is the ability to load, monitor, back up, and optimize the use of the large data tables in the RDBMS. DARPA's Topological Data Analysis program seeks the fundamental structure of massive data sets and in 2008 the technology went public with the launch of a company called Ayasdi. The practitioners of big data analytics processes are generally hostile to slower shared storage,[42] preferring direct-attached storage (DAS) in its various forms from solid state drive (SSD) to high capacity SATA disk buried inside parallel processing nodes. The perception of shared storage architectures—Storage area network (SAN) and Network-attached storage (NAS) —is that they are relatively slow, complex, and expensive. These qualities are not consistent with big data analytics systems that thrive on system performance, commodity infrastructure, and low cost. Real or near-real time information delivery is one of the defining characteristics of big data analytics. Latency is therefore avoided whenever and wherever possible. Data in memory is good—data on spinning disk at the other end of a FC SAN connection is not. The cost of a SAN at the scale needed for analytics applications is very much higher than other storage techniques. There are advantages as well as disadvantages to shared storage in big data analytics, but big data analytics practitioners as of 2011 did not favor it.

Big data has increased the demand of information management specialists in that Software AG, Oracle Corporation, IBM, Microsoft, SAP,EMC, HP and Dell have spent more than $15 billion on software firms specializing in data management and analytics. In 2010, this industry was worth more than $100 billion and was growing at almost 10 percent a year: about twice as fast as the software business as a whole. Developed economies make increasing use of data-intensive technologies. There are 4.6 billion mobile-phone subscriptions worldwide and between 1 billion and 2 billion people accessing the internet. Between 1990 and 2005, more than 1 billion people worldwide entered the middle class which means more and more people who gain money will become more literate which in turn leads to information growth. The world's effective capacity to exchange information through telecommunication networks was 281 petabytes in 1986, 471 petabytes in 1993, 2.2 exabytes in 2000, 65 exabytes in and it is predicted that the amount of traffic flowing over the internet will reach 667 exabytes annually by 2014. It is estimated that one third of the globally stored information is in the form of alphanumeric text and still image data is the format most useful for most big data applications. This also shows the potential of yet unused data (i.e. in the form of video and audio content).

While many vendors offer off-the-shelf solutions for Big Data, experts recommend the development of in-house solutions custom-tailored to solve the company's problem at hand if the company has sufficient technical capabilities. There are many tools and products that support Bigdata like Amazon EMRFS, Hadoop HDFS, IBM BigInsights with supporting development environments with MapReduce, scala, yarn, hive, pig, scoop, zookeeper etc. Some of them are native environments with easy development and deployment on various databases and no-sql environments. No-sql environments supports content management.

16.10 Big Data Case Studies

Retail

Walmart handles more than 1 million customer transactions every hour, which are imported into databases estimated to contain more than 2.5 petabytes (2560 terabytes) of data – the equivalent of 167 times the information contained in all the books in the US Library of Congress.

Retail Banking

FICO Card Detection System protects accounts world-wide. The volume of business data worldwide, across all companies, doubles every 1.2 years, according to estimates.

Real Estate

Windermere Real Estate uses anonymous GPS signals from nearly 100 million drivers to help new home buyers determine their typical drive times to and from work throughout various times of the day

Science

The Large Hadron Collider experiments represent about 150 million sensors delivering data 40 million times per second. There are nearly 600 million collisions per second. \

Science and Research

When the Sloan Digital Sky Survey (SDSS) began collecting astronomical data in 2000, it amassed more in its first few weeks than all data collected in the history of astronomy. Continuing at a rate of about 200 GB per night, SDSS has amassed more than 140 terabytes of information. When the Large Synoptic Survey Telescope, successor to SDSS, comes online in 2016 it is anticipated to acquire that amount of data every five days.

Decoding the human genome originally took 10 years to process, now it can be achieved in less than a day: the DNA sequencers have divided the sequencing cost by 10,000 in the last ten years, which is 100 times cheaper than the reduction in cost predicted by Moore's Law The NASA Center for Climate Simulation (NCCS) stores 32 petabytes of climate observations and simulations on the Discover supercomputing cluster

16.11 Bigdata Execution

Big data analysis is often shallow compared to analysis of smaller data sets. In many big data projects, there is no large data analysis happening, but the challenge is the extract, transform, load part of data preprocessing. Big data is a buzzword and a "vague term", but at the same time an "obsession" with entrepreneurs, consultants, scientists and the media. Big data showcases such as Google Flu Trends failed to deliver good predictions in recent years, overstating the flu outbreaks by a factor of two. Similarly, Academy awards and election predictions solely based on Twitter were more often off than on target. Big data often poses the same challenges as small data; and adding more data does not solve problems of bias, but may emphasize other problems. In particular data sources such as Twitter are not representative of the overall population, and results drawn from such sources may then lead to wrong conclusions. Google Translate - which is based on big data statistical analysis of text - does a remarkably good job at translating web pages. However, results from specialized domains may be dramatically skewed. On the other hand, big data may also introduce new problems, such as the multiple comparisons problem: simultaneously testing a large set of hypotheses is likely to produce many false results that mistakenly appear to be significant. Ioannidis argued that "most published research findings are false" due to essentially the same effect: when many scientific teams and researchers each perform many experiments (i.e. process a big amount of scientific data; although not with big data technology), the likelihood of a "significant" result being actually false grows fast - even more so, when only positive results are published.

- Easy and quick absorption of changes to solution
- Automation of change management solution
- Enabling business users to realize real-time reporting enhancements

16.12 BigData and BI

Big data is growing fast as organizations devote technology resources to tapping the terabytes (if not petabytes) of data flowing into their organizations and externally in social media data and other sources. What does this all mean for business intelligence (BI) users and systems? With all the attention on advanced analytics for big data, what's the play for BI?

Integrating advanced analytics for big data with BI systems is an important step toward gaining full return on investment. Advanced analytics and BI can be highly complementary; advanced analytics can provide the deeper, exploratory perspective on the data, while BI systems provide a more structured user experience. BI systems' richness in dashboard visualization, reporting, performance management metrics, and more can be vital to making advanced analytics actionable.

Table 16.12.1 - Comparison of Data generation techniques in existing big data bench marks

Benchmark Efforts	Volume	Velocity	Variety (data sources)	Veracity
Hi Bench	Partially Scalable	Un-controllable	Texts	Un-Considered

Grid-mix	Scalable	Un-controllable	Texts	Un-Considered
Pig-mix	Scalable	Un-controllable	Texts	Un-Considered
YCSB	Scalable	Un-controllable	Texts	Un-Considered
Sig-MOD	Scalable	Un-controllable	Texts, Tables	Un-Considered
TPC-DS	Scalable	Semi-controllable	Texts	Partially Considered
Bigdatabench	Scalable	Semi-controllable	Texts, Graphs, Tables	Considered

16.12.2   *Bench Marking Techniques*

Benchmark efforts	Work loads		Software Stack	Use-case
	Types	Examples		
HiBench	Off-line Analytics/Real Time Analytics	Sort, world count, Tera sort, page rank, k-mean, bias classification Notch indexing	Hadoop and Hive	Hadoop
GridMix	On-line services	Sort, sampling a large dataset	Hadoop	Hadoop
PigMix	On-line services	12 data queries	Hadoop	Hadoop
YCSB	On-line service	OLTP (read write scan update	No-Sql systems	No-sql and sql systems

SIGMOD	On-line service	Count url links	DBMS and Hadoop	DBMS
TPC-DS	On-line service	Data loading	DBMS	DBMS and Hadoop
Big Data Bench	On-line service/Real Time analytics	Database operations	No-sql DB	Distributed systems

Table 16.12.3 - Bigdata Analytics- Profiling the use of Analytical platforms in User Organizations

TECHNOLOGY	DESCRIPTION	VENDOR/PRODUCT
Massively parallel processing analytic databases	Row-based databases designed to scale out on a cluster of commodity servers and run complex queries in parallel against large volumes of data	Teradata Active Data Warehouse, Greenplum (EMC), Microsoft Parallel Data Warehouse, Aster Data (Teradata), Kognitio, Dataupia
Columnar databases	Database management systems that store data in columns, not rows, and support high data compression ratios.	ParAccel, Infobright, Sand Technology, Sybase IQ (SAP), Vertica (Hewlett-Packard), 1010data, Exasol, Calpont
Analytical appliances	Preconfigured hardware-software systems designed for query processing and analytics that require little tuning.	Netezza (IBM), Teradata Appliances, Oracle Exadata, Greenplum Data Computing Appliance (EMC)
Analytical bundles	Predefined hardware and software configurations that are certified to meet specific performance criteria, but customers must purchase and configure themselves.	IBM SmartAnalytics, Microsoft FastTrack
In-memory databases	Systems that load data into memory to execute complex queries.	SAP HANA, Cognos TM1 (IBM), QlikView, Membase
Distributed file based systems	Distributed file systems designed for storing, indexing, manipulating and querying large volumes of unstructured and semi-structured data.	Hadoop (Apache, Cloudera, MapR, IBM, HortonWorks), Apache Hive, Apache Pig

Analytical Services	Analytical platforms delivered as hosted or public-cloud-based services.	1010data, Kognitio
Non-relational	Non-relational databases optimized for querying unstructured data as well as structured data.	MarkLogic Server, MongoDB, Splunk, Attivio, Endeca, Apache Cassandra, Apache Hbase
CEP/streaming engines	Ingest, filter, calculate and correlate large volumes of discrete events and apply rules that trigger alerts when conditions are met.	IBM, Tibco, Streambase, Sybase (Aleri), Opalma, Vitria

16.17 Data Lakes

A data lake is a method of storing data with in a system or repository, in its natural format, that facilitates the collocation of data in various schemata and structural forms, The idea of data lake is to have a single store of all data in the enterprise ranging from raw data to transformed data which is used for various tasks including **reporting**, **visualization**, analytics. The data lake includes structured data from relational databases and semi-structured data using a centralized data store accommodating all forms of data

Pattern based modelling and use cases.

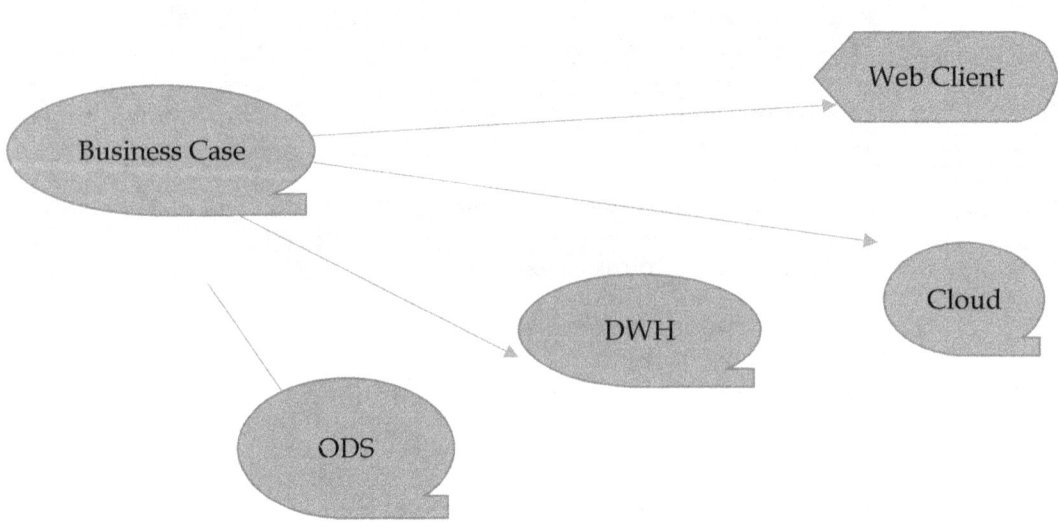

16.18 Data Science

Data science, also known as data-driven science, is an interdisciplinary field about scientific methods, processes, and systems to extract knowledge or insights from data in various forms, either structured or unstructured, similar to data mining.
Some of the basic Data Science Models

- Data Randomized Algorithms. The primary topics in this part of the specialization are: asymptotic ("Big-oh") notation, sorting and searching, divide and conquer (master method, integer and matrix multiplication, closest pair), and randomized algorithms (QuickSort, contraction algorithm for min cuts).
- Statistics with R
- Applied Data Science with Python

Example Data Science Framework

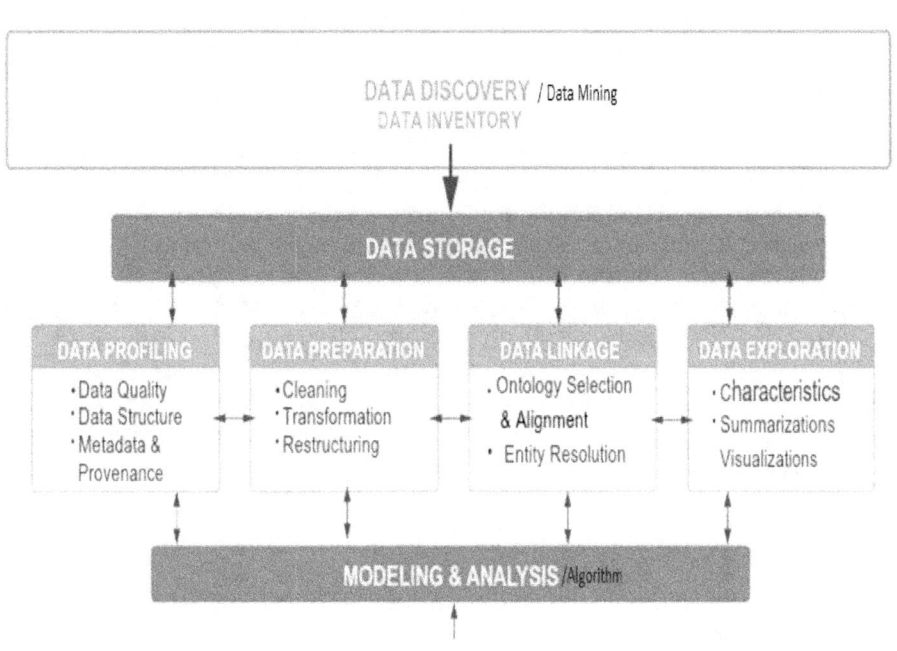

DATA DISCOVERY / Data Mining
DATA INVENTORY

DATA STORAGE

DATA PROFILING
• Data Quality
• Data Structure
• Metadata & Provenance

DATA PREPARATION
• Cleaning
• Transformation
• Restructuring

DATA LINKAGE
. Ontology Selection & Alignment
• Entity Resolution

DATA EXPLORATION
• Characteristics
• Summarizations
Visualizations

MODELING & ANALYSIS /Algorithm

CHAPTER 17
ARTIFICIAL INTELLIGENCE, NEURAL NETWORKS FOR BI AND DSS

Neural Networks is the working paradigm of human neuron, works on stimulus from the neural, from outside the world. Neural networks using the artificial intelligence are named as Artificial Neural Networks. An AIN consists of many single processors to interact through a dense web interactions. A neuron or processing element has primarily will do it computes output and it updates local memory, i.e. weights and other types of data called data variables. Backpropagation calculates the difference between the expected and actual output value. The advantage of ANN is parallelization and predictive support for fault tolerance.

17.1 Accounting

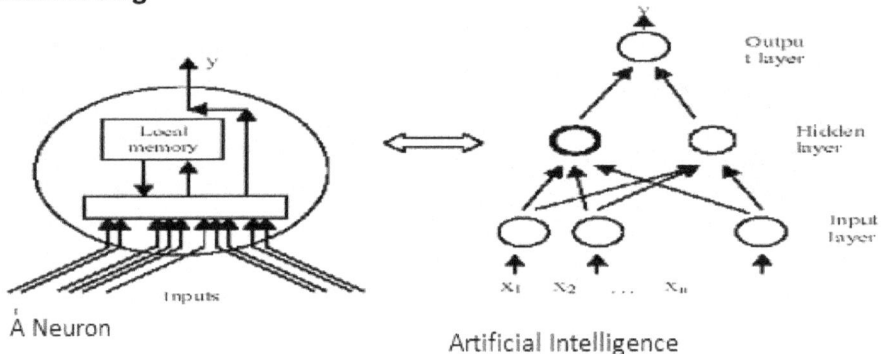

Figure 17.1 – A Neuron and Artificial Intelligence Neural Network Framework

17.2 Business Applications of Artificial Neural Networks

Artificial Neural Networks or ANN has a multitude of real world applications in the business domain which have been classified as follows:
* Identifying tax fraud
* Enhancing auditing by finding irregularities

Finance

* Foreign exchange rate forecasting
* Country risk rating
* Predicting stock initial public offerings
* Stock Market Prediction
* Bankruptcy prediction
* Customer credit scoring
* Risk management

Marketing
* Classification of consumer spending patterns

17.3 AI Bankruptcy Prediction System Framework

The structural and statistical recommended pattern AIN parameters based on modeling the underlying dynamics of interest rates and firm characteristics and deriving the default probability based on these dynamics and instead of modeling the relationship of default with the characteristics of a firm, this information can be derived from the input layer, output layer and hidden layer of data.

The parameters used for identifying the risk:
1) Working capital/total assets
2) Retained earnings/total assets
3) Market capitalization/total debt
4) Total assets vs Conformed

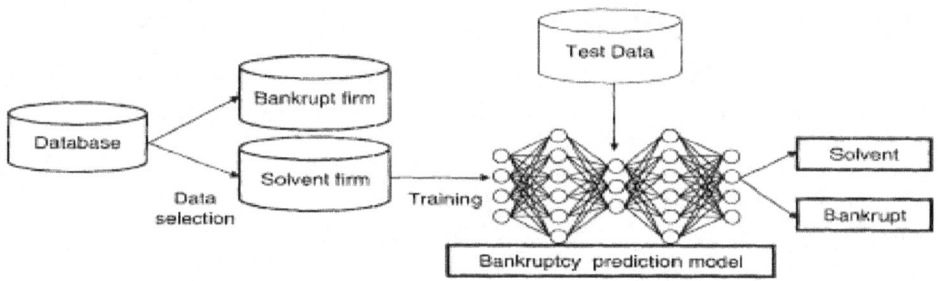

Figure 17.2– Artificial Intelligence Bankruptcy Prediction System Framework

17.4 Applications of ANN in Stock Market Prediction

There are many advocated proposal deployed in an AI based neural network for stock prediction. This system uses sufficient amount of historical stock data as input and then train the network with this data. Once trained the neural network can be used to predict stock behavior. There are various types stock prediction AIN BI models one is frequently used applications gives more information, NN's for stock price predictions which supports systems try to predict stock prices for one or more days in advance, based previous attempts of trained neural networks. The output of the system is the decision to buy and sell and is mostly based on fuzzy logic where by system gives the decision not as a binary signal but as fuzzy signal with a certain percentage of success.

S. No	Problem Domain	Solutions Given
1	Predicting stock performance	Backpropagation/Boltzmann Machine/ELM
2	Stock Price Predictions	Backpropagation Perceptron, Adalina/Madaline
3	Modeling the stock performance (ANN combined)	Backpropagation Hybrid approach (Backpropagation NN + Expert system)/ELM

Table 17.1 Solutions provided in Stock Prediction Market

Books and References

1. W.H Inman – Building the Data Warehouses- Timely Practical Reliable – Third edition – Willey Publishers Culprint
2. Ralph Kimball The Data Warehouse Lifecycle Toolkit, Second Edition: Practical Techniques for Building Data Warehouse and Business Intelligence Systems
3. Somina Venkata Surya brahma Linga Sarma, "Information Management and Accessibly in a cloud vs Convectional Business Intelligence", white paper presented at BIDW 2010 Annual International Academic Conference on Business Intelligence and Data Warehousing (BIDW 2010)
4. Somina Venkata Surya Brahma Linga Sarma -UST International --'Top Seven Most significant latest trends in Data Warehousing and Business Intelligence' June 2008, UST International
5. Somina Venkata Surya Brahma Linga Sarma - CRM Analytics – 'Advancements and Analysis on Customer Relationship Management Analytics' July 2005, Tech Mahindra Satyam
6. Somina Venkata Surya Brahma Linga Sarma- "Effective Implementation of Advanced Analytics using Netezza DWA" was selected for presentation at Business Intelligence Asia Pacific Summit at Pan Pacific, Singapore in Sept 2012
7. Somina Venkata Surya Brahma Linga Sarma – "Scalability and Operational Metrics of various Database Analytics Engines" presented at the 6TH ANNUAL international conference on ICT- Big Data, Cloud and Security (ICT-CS 2015) held in July 27-28 2015
8. Data Growth Challenges Demand Proactive Data Management by Merv Adrian, Principal, IT Market Strategy www.itmarketstrategy.com
9. Technical White Paper Ubuntu Enterprise Cloud Architecture By Simon Wardley, Etienne Goyer & Nick Barcet – August 2009
10. IBM Netezza Data Warehouse Appliance Release 6.0 Highlights, @2010 Winter Corporation Cambridge, MA, USA.
11. Netezza: Enabling Advanced Analytics, @2010 Bloor Research
12. Data Stores, Data Warehousing, and the Zachman Framework: Managing Enterprise Knowledge - Kevin McDonald, a consulting organization specializing in data warehousing and SAP
13. Open Source Data Warehousing and Business Intelligence, CORD project
14. Map Reduce: Simplified Data Processing on Large Clusters - Jeffrey Dean and Sanjay Ghemawat – Google Inc
15. Data In the Cloud: The Changing Nature of Managing Data Accessibility Gartner RAS Core Research Note G00165291, Eric Thoo, 27 February 2009, RA2 12302009
16. Gartner Says Emerging Technologies Will marginalize its Role in Business Intelligence, Gartner, March 2008
17. Artificial Neural Networks: The next intelligence By Amit Khajanchi
18. Enterprise Architecture and Cloud Computing, 2009 Booz Allen Hamilton
19. Using Downstream Data to Create Integrated Business Performance, AMR Research, 2007 by Hussain Mooraj, Lora Cecere, and Chris Fletcher
20. Pacific Asia Conference on Information Systems (PACIS) PACIS 2004 Proceedings, Applying Data Mining to Telecom Churn Management Shin-Yuan Hsiu-Yu
21. Data Warehousing in the Age of Big Data (The Morgan Kaufmann Series) on Business Intelligence – Krish Krishnan – ELSEVIER print
22. Mastering the SAP Business Information Warehouse - Richard M Dunning – American SAP user group _ Wiley Publisher
23. Adjoined Dimension Column Clustering (ADC Clustering) to improve Data Warehouse Query Performance Computer Science Department, University of Massachusetts Boston

24. Data Growth Challenges Demand Proactive Data Management , Principal, IT Market Strategy www.itmarketstrategy.com

ABOUT THE AUTHOR

Venkata Surya Brahma Linga Sarma Somina is acknowledged and honored as Senior Principal Architect of Data Warehousing and Business Intelligence Technologies and Solutions a partner in TDWI and GSTF. He has presented many white papers on Data Warehousing and Business Intelligence solutions and technologies and also frequent visitor of the conferences held by the Global Science and Technology Forum, Singapore. Mr. Sarma has over 18 years of experience in IT industry and has experience of architectural design and development of databases, data warehousing, and business intelligence related application architect, projects encompassing major industry domains such as BFSI, Health Care, telecommunications, retail for clients across the United States, UK, and Asia. He is well-versed in the primary Oracle, Netezza and Teradata databases and in-built database analytics platforms and cloud computing and technology adoption and modernization of data. He has certified in Oracle, Informatica, Netezza and CSQA. He has published many papers with-in organizations to help the business. Mr. Sarma completed his Master of Technology in Process Metallurgy from IIT, Bombay in 1997. Mr.Sarma has expertise in data modeling with conventional and pattern based modelling and design of enterprise data warehousing/business intelligence information with MPP and Shared Nothing Architectures and has pronounced experience implementing solutions using Zachman Frameworks, Open Group Frame Works and Downstream implementations.

Publications

- Tech Mahindra Satyam-CRM Analytics – 'Advancements and Analysis on Customer Relationship Management Analytics' July 2005
- UST International --'Top Seven Most significant latest trends in Data Warehousing and Business Intelligence' June 2008, UST International – won best research paper award
- DWBI2010-'Information Management and Accessibility in a cloud vs conventional business intelligence' presented at international conference DWBI2010 held in Singapore during July 2010 conducted by GSTF
- BI Asia Pacific Summit –'Effective Implementation of Advanced Analytics using Netezza DWA' at Business Intelligence Asia Pacific Summit at Pan Pacific, Singapore in September 2012 conducted by GSTF
- ICT-CS 2015- 'Scalability and Operational Metrics of various Database Analytics Engines' selected for presentation at the 6th annual international conference on ICT-Big Data, Cloud and Security (ICT- CS 2015) held in July 27-28 2015 conducted by GSTF

www.ingramcontent.com/pod-product-compliance
Lightning Source LLC
Chambersburg PA
CBHW080819180526
45168CB00006B/2504